Numanities - Arts and Humanities in Progress

Volume 19

The series originates from the need to create a more proactive platform in the form of monographs and edited volumes in thematic collections, to contribute to the new emerging fields within art and humanistic research, and also to discuss the ongoing crisis of the humanities and its possible solutions, in a spirit that should be both critical and self-critical.

"Numanities" (New Humanities) aim at unifying the various approaches and potentials of arts and humanities in the context, dynamics and problems of current societies.

The series, indexed in Scopus, is intended to target an academic audience interested in the following areas:

- Traditional fields of humanities whose research paths are focused on issues of current concern;
- New fields of humanities emerged to meet the demands of societal changes;
- Multi/Inter/Cross/Transdisciplinary dialogues between humanities and social and/or natural sciences;
- Humanities "in disguise", that is, those fields (currently belonging to other spheres), that remain rooted in a humanistic vision of the world;
- Forms of investigations and reflections, in which the humanities monitor and critically assess their scientific status and social condition;
- Forms of research animated by creative and innovative humanities-based approaches;
- Applied humanities.

More information about this series at http://www.springer.com/series/14105

Simona Stano · Amy Bentley
Editors

Food for Thought

Nourishment, Culture, Meaning

Springer

Editors
Simona Stano
University of Turin
Turin, Italy

Amy Bentley
New York University
New York, NY, USA

ISSN 2510-442X ISSN 2510-4438 (electronic)
Numanities - Arts and Humanities in Progress
ISBN 978-3-030-81114-3 ISBN 978-3-030-81115-0 (eBook)
https://doi.org/10.1007/978-3-030-81115-0

This Springer imprint is published by the registered company Springer Nature Switzerland AG
The registered company address is: Gewerbestrasse 11, 6330 Cham, Switzerland

Foreword: Good to Think With

This compelling collection of essays draws on a conference held at New York University in October 2019. The conference, co-organised by Simona Stano, a faculty member in the Department of Philosophy and Educational Sciences, University of Turin, and Amy Bentley, professor of food studies at New York University, convened scholars from around the world, giving them space and time to share ideas exploring the cultural meanings and uses of food from a variety of perspectives, often over good food.

Each unit in this volume unpacks a theoretical and empirical problem. It opens with a general model of how something so subjective as food is connected to other peoples' practices and thinking. This is done through language, linking two forms of orality; food is what goes in and language is what comes out. Both have a grammar. It is the rules of cooking, serving and eating, and modes of meaning-making that this volume is after. A subset of signification is zeroed in on, in an early chapter, where food is sometimes paradoxically given a face, when Apulian bakers construct marzipan cakes in the shape of fish and lamb, while real fish heads and lamb heads are obscured in dishes. What is the temptation and the taboo here to give food a face? In a sense, this book is about the very science of intersubjective communication and obfuscation and the constitutive signs of such communication, and its limitations.

A second cluster of chapters engages with the problem of heritage-making in two directions: it turns our face away from the exclusive focus on European terroir strategies (which when uncritically celebrated turn out to be forms of ethnic cleansing about roots, excluding those with more portable practices) in attending to actors in Jeddah, Saudi Arabia, and in Tripoli, Lebanon, and secondly focus on heritage as a view of the past from the present. As the present changes, so does its relationship to the past, which while opening up some modes of meaning-making, closes others. Heritage is what current groups of unevenly endowed classes of actors fight to assert and assemble as their past.

Law is the register of social struggles fought, lost and won. A 1929 Missouri General Assembly bill called for the "sterilisation of persons convicted of murder, rape, chicken stealing…". The author asks what is chicken stealing doing in that list of capital crimes and what might it reveal about the relationship between race-making and law-making in American history? If that is an exemplar of state-sponsored violence for trivial crimes matched by deadly and disproportionate punishment, another author turns the question around. She takes two examples of the Black Panther Party's 1969 innovative free school breakfast programme and the I-Collective's struggle to give voice to Black and indigenous food activists. She reveals the relationship between the medium and the message on the new terrain of food-related social media in the pursuit of social justice. In the process, she shows us how the new social media may not merely be the machine for the production of celebrity self-indulgence.

Modern life is unimaginable without advertising and media, which underlines the question about the relationship between reality and representation, undermining the easy fixity of either. Taking the example of advertising for tripe over the last hundred years in the UK, the authors outline the attempt to symbolically represent a sense of hygiene, purity, economy and self-esteem that was eventually subverted by cheap muscle meat and the social construction of disdain, disgust and symbolic violence organised around class and locality. In another lovely chapter, the author poses the acute question: what happens when food becomes the plaything of the young—for instance on Instagram, as the audience for Netflix Master Chef, and via YouTube binge-eating shows—rather than the domain of the more mature cook and care-giver? In the process, this chapter seeks to provide much needed order to the vast landscape of food media in the twentieth-century and its weird proliferation in the already over-burdened twenty-first century.

A range of scholars in the final section argue that the current gastromania around food consumption, materially and symbolically, necessitates the extension of the analytical aperture beyond need and nutrition. Starting from Barthes, where food is "a system of communication, a body of images, a protocol of usages, situations and behaviours", these authors push the semiotic envelop to open up to aesthetic, ludic, critical and utopian possibilities. From superfoods that promise to cure us of the malaise of modernity to new definitions for the toxicity of sugar, salt and fat, the range of fads demands new critical extensions to nutritionism, the gender of diets and orthorexia.

The collection as a whole provides subtle instances of the bridge and the barrier between the textual and the paratexual, the subject and the object, reality and representation, individual ideas in our heads and collective discourses, and everyday practices to big theories of the self and the other. It brings to a head the conflict between science and everyday experience, pressing on the subject, the object and the intersubjective constitution of knowledge about our bodies, its ills and potential cures, both as myth and as medicine. This lively collection draws our attention to the gap between thinking and doing, where thinking is also a form of doing. It successfully connects micro-analysis and close-reading, to macro-changes and distant speculations, without leaching the thing, the sign and the subject of everything that might be ensconced within it. It takes us successfully from semiotic instance to broader theoretical hypotheses, about something as quotidian and ubiquitous as food, which at it turns out, is indeed very good to think with.

Krishnendu Ray
Chair of Department of Nutrition
and Food Studies
New York University
New York, USA

Acknowledgments

This project has received funding from the European Union's Horizon 2020 research and innovation programme under the Marie Sklodowska-Curie grant agreement No. 795025.

Its realisation was possible thanks to the kind support of the Department of Philosophy and Educational Sciences of the University of Turin, Italy, and the Food Studies Programme of New York University, Steinhardt, USA.

We are also very thankful to all the academic centres and other entities that allowed the organisation and fruitful realisation of the international conference *Food for Thought* (New York, October 14–15, 2019), as well as to the scholars and public who very actively participated in it, originating a profitable and stimulating debate, which paved the way for this book.

Contents

Editors and Contributors

About the Editors

Simona Stano is Tenure-Track Assistant Professor at the University of Turin (UNITO, Italy) and Visiting Researcher at New York University (NYU, USA). She has been awarded a Marie Curie Global Fellowship for a research project (COMFECTION, 2019–2021) on the semiotic analysis of food communication. She also worked as Senior Researcher at the International Semiotics Institute (2015–2018) and as Visiting Researcher at the University of Toronto (2013), the University of Barcelona (2015–2016) and Observatorio de la Alimentación (2015–2016).

She deals mainly with semiotics of culture, food semiotics, body semiotics and communication studies and has published several papers, edited volumes (including special issues of top semiotic journals such as *Semiotica* and *Lexia*) and monographs (*I sensi del cibo*, 2018; *Eating the Other. Translations of the Culinary Code*, 2015) on these topics. Since 2020, she is the vice-Director of the Centre for Interdisciplinary Research on Communication (CIRCe) of the University of Turin.
e-mail: simona.stano@unito.it

Amy Bentley is Professor in the Department of Nutrition and Food Studies at New York University. As historian with interests in the social, historical and cultural contexts of food, she is the author of *Inventing Baby Food: Taste, Health, and the Industrialization of the American Diet* (University of California Press, 2014), *Eating for Victory: Food Rationing and the Politics of Domesticity* (University of Illinois, 1998), editor of *A Cultural History of Food in the Modern Era* (Berg, 2011), as well as articles on such diverse topics as ketchup in Reagan's America, the politics of southwestern cuisine and a historiography of food riots. She is also involved in a number of food-related academic and applied projects, including as the co-founder of the NYU Urban Farm Lab and the Experimental Cuisine Collective (2007–2016). She is the former editor-in-chief of *Food, Culture, and Society: An International*

Journal of Multidisciplinary Research (2013–2019) and is the co-editor of the book series *Food in Modern History: Traditions and Innovations* (Bloomsbury).
e-mail: amy.bentley@nyu.edu

Contributors

David Michael Bell College of Arts and Sciences, Ohio University, Athens, USA

Amy Bentley Department of Nutrition and Food Studies, New York University, New York, USA

Emily J. H. Contois Department of Media Studies, The University of Tulsa, Tulsa, OK, USA

Suzanne Cope Expository Writing Program, New York University, New York, USA

Neve Durrwachter Department of Anthropology and Sociology, Ursinus College, Collegeville, PA, USA

Cristina Greco Advertising Communication, Kingdom of Saudi Arabia, University of Business and Technology, Jeddah, Saudi Arabia

Gareen Hamalian Department of Psychiatry, New York University, New York, NY, USA

Massimo Leone Department of Chinese Language and Literature, University of Shanghai, Shanghai, China;
Department of Philosophy and Education Sciences, University of Turin, Turin, Italy

Theresa Moran College of Arts and Sciences, Ohio University, Athens, USA

Henry Peck Faculty of Oriental Studies, University of Oxford, Oxford, UK

Franciscu Sedda Department of Literature, Languages, and Cultural Heritage, University of Cagliari, Cagliari, Italy

Simona Stano Department of Philosophy and Educational Sciences, University of Turin, Turin, Italy

Bruno Surace Department of Philosophy and Educational Sciences, University of Turin, Turin, Italy

Daniel Thoennessen Department of Nutrition and Food Studies, New York University, New York, USA

Ugo Volli Department of Philosophy and Educational Sciences, University of Turin, Turin, Italy

Lauren A. Wynne Department of Anthropology and Sociology, Ursinus College, Collegeville, PA, USA

Chapter 1
Food for Thought: An Introduction

Simona Stano and Amy Bentley

Abstract This book proposes a collective reflection on the relation between *food* (nourishment) and *thought* (culture and meaning), calling into action various theoretical approaches and analytical methodologies, and also offers new insights on how the study of food can help us understand better what we call "culture". The work is structured into three main sections: *Food, Taste, and Global Cultures; Law, Power, and Media; and Nutrition and Culture.* The essays that follow also encompass a series of crucial methodological issues concerning the study of food as a bearer of senses and sense, of value and culture, allowing leading scholars and young researchers from different fields (including semiotics, anthropology, sociology, history of food and media studies) to engage in a fruitful interdisciplinary dialogue.

Keywords Food studies · Semiotics · Culture · Materiality · Meaning

Elected best pastry chef by the Relais Desserts Association (2016), Les Grandes Tables du Monde (2017), Gault&Millau (2018) and The World's Best 50 Restaurants (2018), Cédric Grolet has also become famous worldwide for his 2 million-follower Instagram account (@cedricgrolet), where he frequently posts pictures of his culinary creations. Among these, his "cube cakes"[1] have attracted a great attention. Made of small, cube-shaped cakes assembled together, Grolet's treats recall the famous Rubik's cube, interestingly pointing to a conception of food as a "puzzle", that is, a "brain-teaser" requiring careful consideration and exploration.

In fact, as Claude Lévi-Strauss (1962) effectively showed, in order to be "good to eat" (*bon à manger*), substances must be first of all "good to think" (*bon à penser*). In other words, food must nourish people's collective mind—i.e. their systems of values,

[1] www.instagram.com/p/B0lSDolIUaB/?utm_source=ig_web_copy_link.

S. Stano (✉)
Department of Philosophy and Educational Sciences, University of Turin, Turin, Italy
e-mail: simona.stano@unito.it

A. Bentley
Department of Nutrition and Food Studies, New York University, New York, USA
e-mail: amy.bentley@nyu.edu

S. Stano and A. Bentley (eds.), *Food for Thought*, Numanities - Arts and Humanities in Progress 19, https://doi.org/10.1007/978-3-030-81115-0_1

beliefs, and traditions—to be considered suitable for their stomachs. Other scholars have since weighed in on the nature of food and culture, including cultural materialists (Harris 1985), who argue that the locus of cultural meaning resides in objects (rather than ideas about the objects) and also practice theorists (Warde 2014, 2016), who assert that a focus on practices and actions provides a third way to think about culture and meaning, sidestepping tensions between emphasis on ideas and things. While food habits, preferences, and taboos are partially regulated by ecological and material factors, research has shown that all food systems are structured and given particular functioning mechanisms by specific societies and cultures, either according to totemic (such as in animistic religions), sacrificial (such as in ancient history), hygienic-rationalist (such as in contemporary Western dietetics), aesthetic (such as in gastronomy), or other types of symbolic logics.

This provides much "food for thought". This commonplace expression assumes an intensified relevance in our era of globally connected food systems as well as instantaneous communications. Not only do cultures develop unique practices for the production, treatment and consumption of food, but such practices inevitably end up affecting also food-related aspects and spheres that are generally perceived as objectively and materially defined. Let us consider, for instance, dietary prescriptions, which are undoubtedly based on the material composition of food products, but are also dependent on the values and meanings conferred on specific food constituents by the narratives and discourses circulating within each culture; or food safety regulations, which are related to the concepts of dirtiness and hygiene—whose perception, as Mary Douglas (1966) effectively showed, is intrinsically related to cultural diversity.

On the other hand, Grolet's cakes reveal a clear attempt to "conceal", at least momentarily, food materiality—i.e. flavours, textures, substances, ...—and meanings—i.e. socio-cultural connotations—under a common "form", following an increasing trend in experimental cuisine, as well as in food design and communication. Let us consider, for instance, the well-known graphic project "Cubes" by Studio Lernert & Sander,[2] portraying ninety-eight different foods cut into 2.5 by 2.5 by 2.5-cm cubes and arranged to form a symmetrical pattern. When asked about the meaning of their work, commissioned by the newspaper *de Volkskrant* for a food-related feature, the designers replied: "Food is an overwhelming subject ... You can go so many different ways. How can you photograph something when you can't decide?" (in Goldberg 2015). Eventually, the solution they opted for was "mak[ing] all of the food seem equally important by cutting everything into uniform pieces" (Ibid.). However, as we all know, such a uniformity cannot but be "ostensible", since unavoidable "differences" make food substances distinct from each other. In fact, not long after the release of Lernert & Sander's picture a "Food Cubes Quiz"[3] became popular on the web. Building on the "hints"—i.e. those elements such as colours and

[2] http://lernertandsander.com/cubes/.

[3] Which would be accompanied by smells, textures, sounds and flavours in a complete food experience extending beyond the purely visual perception offered by such a text.

structural configurations that cannot be totally concealed—suggested by the cubes portrayed in the designers' picture, players were requested to recognise the ninety-eight food substances depicted in the image, in a certainly difficult—but not at all impossible—interpretative "bet".

Even more interestingly, food substances acquire specific meanings in specific cultures and societies, precisely by virtue of the dynamics described by Lévi-Strauss and the other scholars mentioned above, evidently recalling the idea of a "puzzle" stressed by Grolet's creations. Hence this book proposes a collective reflection on the relation between *food* (nourishment) and *thought* (culture and meaning), calling into action various theoretical approaches and analytical methodologies, also in the aim to offer new insights on how the study of food can help us understand better what we call "culture".

To this purpose, the work is structured into three main sections:

I. Food, Taste, and Global Cultures

Food and taste have always represented crucial means of construction and expression of sociocultural identity, as foundational food studies scholars Lévi-Strauss (1958, 1964, 1965), Barthes (1961), Douglas (1966, 1972, 1984), Bourdieu (1979) have effectively pointed out. What is more, in contemporary societies, migrations, travels, and communications incessantly expose local food identities to global food alterities, originating remarkable processes of transformation that continuously reshape and redefine such identities and alterities (Stano 2015, 2016). This originates a series of interesting questions: how can the cultural meanings and values associated with food be identified and described? How do the processes of hybridization (and domestication) of food and taste affect such meanings and values in today's world?

In the aim to explore these issues, the section opens with Ugo Volli's thought-provoking reflections upon the ceremonial aspects of food consumption, which point to the idea of "alimentation" and the potential of the semiotic approach toward its analysis. A compelling chapter by Massimo Leone follows, offering insights on the effects of meaning of the processes of attribution, or rather concealment, of a face to/from food, and recalling crucial issues in the study of food cultures and rituals, such as the idea of sacrifice and the consumption of non-human animal meat.

With Franciscu Sedda's "Phenomenology of a Symbolic Dish" the attention moves to cultural heritage and identity, encompassing the central problem of (un)translatability and emphasising the role of food in the processes of identity recognition and remembrance. Cultural memory and heritage are also crucial to Cristina Greco's analysis of Saudi "traditional" and "authentic" food, especially in the frame of the ongoing development plan Saudi Vision 2030, and to Henry Peck's study of Tripoli's sweets craft and its symbolic meanings across time and through sociocultural changes.

II. Law, Power, and Media

Both at the local and global scale, nutrition is ruled by a complexity of laws regulating quite diverse aspects—e.g. quality, safety, ecology, etc.—related to the production, trade and handling of food. Such aspects, just as any other facet of law, cannot be disentangled from culture (see in particular Geertz 1983; Rosen 2006). This explains the difficulty that might be encountered in establishing transnational guidelines on food, as well as protocols and procedures on a more local scale. This section focuses on the cultural conceptions underlying food regulations and the way by which they contribute to activate specific meaning-making processes. Furthermore, it highlights the strong link between food and power, conceiving the former as "a weapon and a blanket, a means of control and of protest" (Avieli 2018: 14), that is to say, a tool for conveying, but also violating, social and cultural norms. This also means paying particular attention to media and communication, and the crucial role they play in shaping food and taste collective imaginaries.

Drawing on these premises, Daniel Thoennessen offers an attentive analysis of American chicken theft laws, recognising them as an expression of racialised political violence toward African Americans, while Suzanne Cope interestingly examines the relation between food, activism and political discourse, comparing the Black Panther Party's Free Breakfast for Children Program to the contemporary indigenous group "I-Collective".

Finally, the section deals with the crucial role played by communication and the media, thanks to David Bell and Theresa Moran's careful reconstruction of the advertising campaigns of United Cattle Products' marketing of tripe to British workers in the 1920s and 1930s and Bruno Surace's captivating analysis of the new generations' representations and valorisations of food products and practices, especially as reflected by the digital media.

III. Nutrition and Culture

Nutrition evidently relies on the material dimension of food, since it makes reference to its physical composition (in terms of nutrients, calories, etc.), but is also strongly influenced by the sociocultural sphere: not only do sociocultural factors such as ethnicity, class, education, and gender affect eating habits, but the very ideas of health, beauty, safety and a series of other concepts playing a crucial role in the definition of dietary regimes are culturally defined. Furthermore, contemporary foodways have increasingly emphasised the connection between nourishment and aesthetics (mainly as a result of a generalised process of aestheticisation of food and taste), as well as the link between nutrition and ethics (as a dominant position supporting meat-free dietary regimes clearly shows). The last section of the book reflects upon such issues, also considering the decisive role played by communication, and especially by the mass and new media, in the establishment of specific collective imaginaries and the association of particular values and meanings to food products, habits, and practices.

In particular, Simona Stano offers a thought-provoking reflection on the relation among nutrition, culture and meaning by means of the analysis of some crucial issues in contemporary foodscapes, including the increasingly important role played by nutritional information in both governmental and media communications, as well as the opposition between genetically modified products and organic food. Nutritionism is also central to Emily Contois' contribution, which interestingly addresses the gendered effects of this ideology, examining the stereotypical representations of salad in diet and wellness media. Finally, Lauren Wynne, Gareen Hamalian and Neve Durrwachter examine the lively debate about orthorexia nervosa, comparing medical literature and narrative accounts by individuals who identify themselves as recovering orthorexics, and showing how it has become a salient disease category without even being formalised yet within the Diagnostic and Statistical Manual of Mental Disorders.

This rapid overview offers only a glimpse of the variegated and multifaceted reflections and analyses gathered in this book, which, exactly as Grolet's cakes, recall the Rubik's cube, in an attempt to "taste" all of its flavours, and above all to answer some of the most urgent questions underlying the "brain-teasing" study of food as a system of signs and sociocultural elements. Although research in this field has been increasing and evidently developing over the last decades, in fact, there are still several open questions, as well as a number of relevant case studies that remain unexplored. While contributing to fill this gap, the essays that follow also aim at encompassing a series of crucial methodological issues concerning the study of food as a bearer of senses and sense, of value and culture, making leading scholars and young researchers from different fields (including semiotics, anthropology, sociology, history of food and media studies) talk to each other in a fruitful interdisciplinary dialogue. Let us turn the page, then, and start solving our cube!

References

Avieli, Nir. 2018. *Food and Power: A Culinary Ethnography of Israel*. Oakland: University of California Press.

Barthes, Roland. 1961. Pour une psychosociologie de l'alimentation contemporaine. *Annales ESC* XVI(5): 977–986.

Bourdieu, Pierre. 1979. *La distinction*. Paris: Éditions de Minuit.

Douglas, Mary. 1966. *Purity and Danger. An Analysis of Concepts of Pollution and Taboo*. London: Routledge and Kegan Paul.

Douglas, Mary. 1972. Deciphering a meal. *Daedalus* 101 (1): 61–81.

Douglas, Mary. 1984. *Food in the Social Order: Studies of Food and Festivities in Three American Communities*. New York: Russell Sage Foundation.

Geertz, Clifford. 1983. Local Knowledge: Fact and Law in Comparative Perspective. In *Local Knowledge: Further Essays in Interpretive Anthropology, 167–234. New York: Basic Books.*

Goldberg, Elyssa. 2015. How These 98 Identical Food Cubes Were Made. https://www.bonappetit.com/entertaining-style/trends-news/article/food-cubes-art. Accessed 27 August 2020.

Harris, Marvin. 1985. *Good to Eat. Riddles of Food and Culture*. New York: Simon and Schuster.

Lévi-Strauss, Claude. 1958. *Anthropologie structurale*. Paris: Plon.

Lévi-Strauss, Claude. 1962. *Le totémisme aujourd'hui*. Paris: PUF.

Lévi-Strauss, Claude. 1964. *Mythologiques I. Le cru et le cuit*. Paris: Plon.

Lévi-Strauss, Claude. 1965. Le triangle culinaire. *L'arc* 26: 19–29.

Rosen, Lawrence. 2006. *Law as Culture: An Invitation*. Princeton, NJ and Oxford, UK: Princeton University Press.

Stano, Simona. 2015. *Eating the Other. Translations of the Culinary Code*. Newcastle-upon-Tyne: Cambridge Scholars Publishing.
Stano, Simona. 2016. Lost in Translation: Food, Identity and Otherness. *Semiotica* 211 (1/4): 81–104.
Warde, Alan. 2014. After Taste: Culture, Consumption and Theories of Practice. *Journal of Consumer Culture* 14 (3): 279–303.
Warde, Alan. 2016. *The Practice of Eating*. Cambridge: Polity.

Simona Stano is Tenure-Track Assistant Professor at the University of Turin (UNITO, Italy) and Visiting Researcher at New York University (NYU, US). She has been awarded a Marie Curie Global Fellowship for a research project (COMFECTION, 2019–2021) on the semiotic analysis of food communication. She also worked as Senior Researcher at the International Semiotics Institute (2015–2018) and as Visiting Researcher at the University of Toronto (2013), the University of Barcelona (2015–2016) and Observatorio de la Alimentación (2015–2016).

She deals mainly with semiotics of culture, food semiotics, semiotics of the body, and communication studies, and has published several papers, edited volumes (including special issues of top semiotic journals such as *Semiotica* and *Lexia*), and monographs (*I sensi del cibo*, 2018; *Eating the Other. Translations of the Culinary Code*, 2015) on these topics. Since 2020, she is the vice-Director of the Centre for Interdisciplinary Research on Communication (CIRCe) of the University of Turin.

Amy Bentley is Professor in the Department of Nutrition and Food Studies at New York University. A historian with interests in the social, historical, and cultural contexts of food, she is the author of *Inventing Baby Food: Taste, Health, and the Industrialization of the American Diet* (University of California Press, 2014), *Eating for Victory: Food Rationing and the Politics of Domesticity* (University of Illinois, 1998), editor of *A Cultural History of Food in the Modern Era* (Berg, 2011), as well as articles on such diverse topics as ketchup in Reagan's America, the politics of southwestern cuisine, and a historiography of food riots. Her current research projects include a history of food in US hospitals, and the meanings and uses of food production in religious communities. In addition to her work as a food historian, she's been involved in a number of food-related academic and applied projects, including as co-founder of the NYU Urban Farm Lab and the Experimental Cuisine Collective (2007–2016). The former Editor-in-Chief of *Food, Culture, and Society: An International Journal of Multidisciplinary Research* (2013–2019), Bentley is co-editor of the book series*Food in Modern History: Traditions and Innovations* (Bloomsbury). She serves as a board member for the journals *Food and Foodways*, and the *Graduate Journal of Food Studies*, and is a Faculty Fellow in Residence at Brittany Hall at NYU.

Part I
Food, Taste, and Global Cultures

Chapter 2
Alimentation: A General Semiotic Model of Socialising Food

Ugo Volli

Abstract Food consumption is one of the *primary needs* of the human animal, as central to our existence as, for example, sleep, sex, and the elimination of digestive waste. In most human societies, however, these last needs are met in a condition of more or less formalised and compulsory *intimacy*, while eating is treated as a *social event*, which reaches its full satisfaction in a public environment. This public dimension requires the overcoming of numerous difficulties, from the link between food and the killing of animals to injustice in food distribution, to the "natural" and thus potentially offensive character of food ingestion. To overcome all these problems, the act of food consumption must be "civilized". This happens in different ways. First, there is the preparation of foods, which are rarely left in their natural state. Food processing usually goes beyond the pure needs of conservation or taste, turning edible matter into "dishes". Second, there is the choice of times, places, gatherings that transform food consumption into an *occasion*. Then there is the *ordering* of eating, its serialisation according to a syntagmatic axis (succession) and a paradigmatic one (choice) that often is arranged in advance by those who prepare food, but actualised by those who consume it. This ordering attributes *the semioticcharacter of a text* to food. But this textualisation is enhanced by the presence of paratextual elements that envelop and further "civilize" the act of eating. These are very different linguistic inserts (for example prayers, blessings, conversations, toasts, performed in particular moments by specific subjects), but also gestural and performative rules obliging both those who serve and who consume food, which are specified in real grammars of eating (etiquette, good manners, etc.). From these prescriptions derives also the necessity of particular devices supporting the activities of food consumption (plates, cutlery, napkins, etc.). All this gives food consumption a ceremonial aspect that makes it the object of study for anthropologists, sociologists, scholars of religions, but above all semiotics, because the stakes of all this "superfluous" activity with respect to simple eating is the social meaning of food. This chapter illustrates the theoretical categories of this semiotic ceremonial approach and analyses some relevant examples.

U. Volli (✉)
Department of Philosophy and Educational Sciences, University of Turin, Turin, Italy

S. Stano and A. Bentley (eds.), *Food for Thought*, Numanities - Arts and Humanities in Progress 19, https://doi.org/10.1007/978-3-030-81115-0_2

Keywords Food · Alimentation · Etiquette · Practices · Table manners · Semiotics

2.1 Alimentation as a Scientific Problem

Academic studies about how people feed themselves focus mainly on the *object* of this process, that is to say, *food*: its history, culture, and geography (Stano 2015); the ritual, religious, or aesthetic prohibitions and obligations that affect it; the tastes and distastes regulating its choice (Stano 2018; Volli 2019); and its "grammar", "generative process" and "meaning" (Volli 2016), which are the specific objects of a semiotics of food.

However, the complex *practices* needed for producing, preparing, distributing, consuming, contextualising, and socialising food are not the same scientific object as food itself—for the same reason why one cannot identify visual art with paintings, fashion with clothes, religion with prayers, and language/discourse with words. Of course, there is always a strong relationship between every social practice and its results. However, the first terms in this series designate some *complex social* activities, while the second ones refer to the *objects that are produced* by them. As social scientists, we must be aware of this difference and be careful about choosing the right tools to investigate both levels.

In this chapter, I will use the term "alimentation" to refer to this *whole process* of choosing, producing, making available, socialising and consuming food in its entire complexity, while studies of "nutrition", "feeding", "cooking", "eating", "manners" and the like usually focus on more narrow aspects. The word "alimentation", though less common in English, has an interesting definition: "the act or process of affording nutriment or nourishment" (Merriam Webster 2020). The aim of this essay is precisely to reflect on such an act or process, extending the borders of common research on food. In fact, many semiotic, historical and sociological studies of food have already covered aspects that go "behind" and "beyond" the mere object of food experiences, addressing the *practices* of consumption and production of food and their social context. Let us consider, for instance, the fair amount of research focusing on etiquette, table manners and food habits in different ecological, economic, social and cultural environments, which touch upon some aspects of what I propose to call the "alimentary process". Such topics, however, have remained rather marginal in food studies, and are generally isolated from each other, lacking a *general framework of analysis*. Indeed, it is easier to find some attempts of analysis on alimentation in disciplines such as microsociology, anthropology, or the history of techniques and inventions. Yet in those cases the focus is usually elsewhere, such as on the "process of civilization" (Elias 1939), the "mythical" content of food, class and gender conflict, or on the diffusion of innovations. By contrast, this chapter intends to provide a tentative general framework of the social process of alimentation as such, paying particular attention to its peculiarities. This attempt will also allow me to suggest some additions to the semiotic terminology itself—and hence to its methodology.

2.2 The Alimentary Text

It is evident that alimentation, in the meaning suggested above, is a complex technical and social process, which involves different temporal, geographic, and economic scales. After illustrating some examples of critical points in this process, I will show a possible general model to understand it, drawing on a rhetorical and semiotic model.

Let us start from the technological chain of food. The preparation of a very simple dish like pasta or a hamburger needs dozens of materials, tools, and technical systems: from main ingredients (e.g. wheat, meat) and condiments (tomato, cheese, herbs, salt), to tools (pots, metal grills, dishes), heat sources (wood, coal, gas, electricity), and material culture surrounding the meal (tablecloths and napkins). All these elements in turn require complex materials and technical systems to be realised and distributed (think of the supply chains, cable and pipe networks, and control systems needed to provide electricity, fire or water). Those of us who live in rich and technologically sophisticated societies usually take all these resources for granted, but the logistical and technical chain of alimentation and the corresponding work is huge, even for the simplest food.

Moreover, food production is not just a matter of logistic chains and complex technical and economic systems. Food is always the product of *social* work. Even behind the simplest dish, there are *cultural* decisions with complex historical and semiotic roots: what kinds of organic substances can be transformed into food and therefore eaten; in what circumstances; after what material and social processes of preparation; what taste combinations are allowed and what are not; with whom, how, what times and ways we can eat, and so on. And, of course, there are different economic and social *strategies* operated by all the actors who consume or provide food playing different roles in such a complex system.

People are generally not aware of all these processes; they ignore the technical chains, and even more the cultural work behind food. Cultural determinations, as well as the physical availability of food substances and materials, seem to them just *natural*. Actually, this same "unconsciousness" can be observed for many other complex social activities—and especially for language. But it is easy to see that the forms of alimentation, not just of food, are highly meaningful for every society. Just think of the choreography of formal banquets, of the ritual meals of various religions, or of the particular rhetoric of *symposia* in ancient Greece. In these cases, as in many others, every detail of cooking, organising meals and beverages in time and space, arranging tools and decorations, discourses and prayers, the proxemics of the table, acquires importance and visibility. Often alimentation (and not just food) works as a complex text, able to make specific meanings, sense effects, and relations arise.

But why does this happen? In order to answer this question, it is useful to describe first of all the specificity of alimentation among other human activities. Food consumption is one of the primary needs of the human animal, more or less at the same level, for example, as sleep, air, sex, and the elimination of digestive waste. In most human societies, however, these needs (excepting air) are generally met in a condition of more or less formalised and compulsory *intimacy*, while eating is almost always treated as a *social event*, reaching its full realisation only in socialised environments (from family dinners to institutional banquets). We can certainly eat on our own if there are circumstances (e.g. work, travels, etc.) requiring it. However, these are generally perceived as somewhat incomplete situations, which satisfy the physiological needs of food consumption, but do not meet its psychological and social necessities. This strong social character of food practices does not only imply the fact that the *ways* and *rules* of eating are established at the level of *culture* (and are never free *individual* choices, nor just *universal* biologic instincts), but also that the practice of eating itself generally requires the existence of a *group of table companions* and includes social *interactions* within it. This *public* dimension is an opportunity for people to represent their social roles, the making of *togetherness*—or, as it is generally referred to, "conviviality"—a basis for any "complete" food experience. It is a basic aspect of the feeding process, as much as the supply of nutrients needed by the body.

However, it must not be forgotten that conviviality requires a number of critical and sometimes embarrassing issues, which are more or less socially neglected and psychologically hidden behind the enjoyment of food, to be overcome, neutralised or removed: the link between meat and dairy foods and animal slaughter and exploitation; injustice in food distribution and hunger; the imbalances of power among diners; the animalistic, "too natural" character of the act of food ingestion, which always threatens to emerge from beneath the veneer of sophisticated etiquette; and the presence of potentially dangerous devices such as knives, forks, spits, fires and also parts of animals or other foods[1] To overcome the implicit and potential violence of meals and foster trust at the table, as well as to remove or at least neutralise the other aspects described above, the act of nutrition must be "civilized", that is to say, disciplined by social rules. This *process of civilization* (Elias 1939; cf. Lévi-Strauss 1968) is realised in every society through different processes:

(1) First of all, the *preparation* of food, which is rarely left in its natural state. Cooking is a cultural activity of domestication of food, as Lévi-Strauss (1964) clearly showed. Mostly, food processing goes far beyond the pure needs of conserving products or arranging flavours. It is a complex and often symbolic process that turns what was just a potentially edible matter into dishes, portions, and courses, presenting them "civilly" and often hiding their "bloody" nature. The animal or vegetable origin of food is often made indistinguishable: dimensions, colours, textures, flavours are radically transformed, and made artificial, except for some foods like fruits. Through such practices, the materials of meals are prepared for the "correct" ways of ingestion, which are regulated by detailed rules, such as the prohibition to show one's teeth while biting food,

[1] In fact, several stories and myths connect food with killing. Think of the Atreides, Macbeth, Cain, Polonius, the last dinner of Don Juan with the statue of the Commendatore, and even Jesus' last supper, as well as to all the tragedies that took place during banquets or around food; or consider the very diffused habit for kings and dictators to have table tasters against the threat of being poisoned. Meals are often displays of strength and wealth, but also moments of weakness, times of enjoyment but also of danger.

or to make noise by eating, all in the attempt to control excessively "animal" behaviours. Furthermore, priorities and obligations on how much and when to eat are generally established, and the tools used to prepare and process foods. First and foremost knives, are "domesticated"—e.g. by rounding their blade, as it happens in Europe since eighteenth century[2]—, or completely eliminated from the table—as in China and Japan. A particular *foodaesthetics* is called into question, also mobilising social mechanisms of synaesthesia. Such aesthetics are culturally established, and regulate the colours, smells, temperatures, and standard dimensions attributed to various edible materials following specific patterns and models of cooking and dressing techniques, which in general are rather long-lived and geographically localised, but above all are *taught* as ancient and "natural" and conceived as the *right* ways of treating edible matter for making food out of it.

(2) Secondly, the *choice of times, places* and *people* transform the act of eating into an *occasion* for diners. People do not eat at any time, anywhere, with any person. There are always *rules* (which can be very different in different societies) regulating the number, timing and specific content of meals. In our culture, for instance, we cannot have breakfast at 4 pm, nor have dinner in the morning. The American breakfast offers different foods than the Italian "prima colazione", although it takes place at the same time of day. Formal dinners have become quite similar throughout the western world, but their timing and structure varies greatly, for example between Britain and Spain. People we have breakfast with (generally, our relatives) are often different from the people we meet for a working lunch. What is eaten, as well as when, where, and with whom, represent interwoven details of the food experience, which combine to establish a precise social meaning.

(3) Moreover, we should consider the particular "order" regulating food consumption within each meal, that is to say, its *serialization* according to a *syntagmatic* (succession) and a *paradigmatic* (choice) axis. Generally, this order is arranged in advance by those who prepare food, but it is *actualised* by those who consume it. Both classes of actors, therefore, assent to the *semiotic* character of a *text* to the meal. There are cultures (e.g. Japan, China, Indonesia) where the paradigmatic organisation prevails and others where the syntagmatic principle rules. In Europe, for instance, the paradigmatic organisation of meals, known as *service à la française*, predominated until the nineteenth century; it consisted in the practice of making various dishes of a meal available *at the same time*, with the diners freely helping themselves from the serving dishes.[3] This contrasted with so-called *service à la russe*, in which dishes were brought to the table sequentially and served individually, portioned by servants (usually

[2] During the Renaissance, it was customary for a particularly honoured gentleman or guest to cut portions of meat in front of the public, using knives and other tools in front of the guests. But this habit was gradually abandoned as the "civilization process" proceeded (Elias 1939; Visser 1992).

[3] Each course included a variety of dishes, all set at the same time at the table. Guests served themselves and their neighbours, following fixed precedence patterns.

at a sideboard in the dining room) before being given to the diner. The number of dishes (or courses) served at a meal *à la russe* has changed over time; but an underlying pattern of service—beginning with soup, then moving through various entrées, then to the roast or game, and then to vegetables (including salads), sweets and coffee—has persisted from the mid-nineteenth century (when this type of service was introduced to France) until now, even if in a reduced form.[4]

In almost all European and American menus of last two centuries,[5] we find a very similar organisation as regards not only the prevalence of the syntagmatic order, but also the choice of foods and general aesthetic structures. This is related to the success and spread of French *cuisine*, which has represented the international "language" of the culinary arts for centuries, although in concurrence with local aesthetics of food.

The *sensory configuration* of such syntagmatic organisations is also interesting, since it functions to actualise culturally recognised oppositions, such as cold/warm, sweet/salt/sour, liquid/solid, meet/vegetables/fruit, "light"/"heavy", spicy/plain. In this sense, it is also worth considering the links between foods and wines, which also reflect hierarchies, as well as perceived oppositions and harmonies. In short, there is a *rhetoric of the senses* that constitutes a *cultural style of taste.* Nevertheless, a process of simplification has imposed itself over time, reducing the influence and prestige of the French model (starting from the linguistic level).

(4) The above-described process of textualisation is enhanced by the systematic presence of paratextual elements that *envelop* and also "civilize" the act of eating. There are very different linguistic inserts (including prayers, blessings, conversations, toasts, which are performed in particular moments by specific subjects), but also gestural and performative rules affecting both those who serve and those who consume food. Let us consider some examples.

Prayers are often used before or after consuming the meal to induce us to recognise that food is not really given by the host who offers it (*horizontal* dimension), but comes from "Above", and for this we owe gratitude. This way, eating becomes a religious act, which is justified by the will of the divinity.

Sacrifices have played the same role in many cultures, being replaced by prayers only in more or less recent times. In fact, the sacrifice is strongly linked to food consumption, either when the idea of "feeding the gods" (such as in the myth

[4] Ranhofer (1894), for instance, suggested this arrangement: "Oysters/2 Soups/Side Dishes (hors d'oeuvre) hot and cold/2 Fish, potatoes/1 Remove/vegetables/1 Entrée, vegetables/1 Entrée, vegetables/1 Entrée, vegetables/1 Punch/1 or 2 Roasts/1 or 2 Colds, salad/1 Hot sweet dessert/1 or 2 Cold sweet des'rts/1 or 2 Ices. Dessert". The "modern dinner" suggested by Emily Post (1950), by contrast, consisted more simply in "Soup or oysters or melon or clams/Fish or entrée/Roast/Salad/Dessert/After-dinner coffee".

[5] See, for instance, the examples available at https://pinoazz.files.wordpress.com/2012/01/menu-storici.pdf; https://www.academiabarilla.it/the-academy/biblioteca-gastronomica-e-menu-storici/; https://gearpatrol.com/2015/02/19/a-look-at-historic-menus-from-around-the-world/; https://thetakeout.com/these-historical-menus-show-how-drastically-the-way-we-1798258043.

of Prometheus in classical Greece) is stressed, or—more commonly—when it allows the believers to express their gratitude for the food they consume, or rather to sublimate the violence (e.g. the killing of an animal) supposed by some products. Similar processes concern the libation of liquids, in particular alcoholic beverages. Moreover, the altars for gods, but even those for dead family members, often contain food, which can assist their journey to the other world or as a form of remembrance for those who are still alive.

Toasts and conversations (often mandatory in different times before, during and after the meal) use the favourable climate of conviviality induced by the meal to strengthen communication between the diners and what they represent (e.g. specific countries in official banquets involving institutional delegates).

Furthermore, civilization concerns the devices allowing—and regulating—food consumption (e.g. plates, cutlery, napkins, etc.), and their relations with people (e.g. how cutlery, napkins, plates and glasses are arranged on the table and used, in which order and in which position the guests should place them while eating or at the end of the meal). An interesting example is the fork (Wilson 2012; Rebora 2013; Casey 2009), which was not used for eating food in Europe during Antiquity and the Middle Ages until the eleventh century. In 1004, princess Maria Argyropoulaina, seventeen-year-old daughter of the Byzantine prince Argiro, married the twenty-year-old Giovanni Orseolo, son of the Doge Pietro II. After the wedding in Constantinople the couple arrived in Venice for the festivities. During a banquet, as everyone ate with their hands, the princess took her food with a two-pronged fork made of gold. This tool was already widespread among the wealthy Byzantines, as it had been popularised by the Empress Theodora, but caused a scandal in Venice, where it was severely disapproved by the clergy, who labelled its use as sinful. It took a while before the use of forks became accepted in Italy, at least in the upper echelons of society, but even then forks were only used to pick food from the common dish in the middle of the table. Starting from the sixteenth century, the new implement spread northwards, first to France, then to Germany and finally to England. "God save me from these little forks", Martin Luther exclaimed in 1518. An English traveller who had encountered forks in Italy in 1608 received nothing but scorn and derision when he recommended their use back home. It may seem only a technical aspect, but cutlery profoundly affected the relationship with food. As Karl Marx (1858) observed in his *Grundrisse,* "the hunger gratified by cooked meat eaten with a knife and fork is a different hunger from that which bolts down raw meat with the aid of hand, nail and tooth" (I, 2). Forks have changed not just *how*, but also *what* people have eaten

All these devices therefore represent in different cultures *systems of reciprocal relationships*, which are often displayed in a very clear manner in their arrangement on the space of the food experience. Not only in formal situations, but also in everyday meals, *spatial grammars*, well established at the cultural level, regulate their disposal and use. The diners and their arrangement around the table are also part of such systems and grammars, which vary across different cultures. The way people arrange themselves to eat influences the cutlery, which in turn has an impact on food preparation and consumption, and vice versa. A very interesting case in this sense is the *triclinium* (τρικλίνιον), that is to say, the couch generally used in Greek and Roman banquets—which would probably seem absolutely uncomfortable to most of us, but was an unavoidable element of formal meals in ancient times. Each couch was sized to accommodate a diner, who reclined on his/her left side on cushions. The *triclinium* was characterised by three sofas on three sides of a low square table, whose surfaces sloped slightly away from the table. Diners generally reclined on these surfaces in a semi-recumbent position. The fourth side of the table was left free, presumably to allow service to the table. From the point of view of cultural exchanges, it is interesting to note that this model of eating accommodation, with diners placed on their side on a couch, is also prescribed for the most formal meal of the Jewish liturgy, the Passover Seder (Volli 2014; Segre 2001). In *Haggadah*—a very formalised text composed of different stories, chanting comments, instructions and warnings, which is read during this liturgical dinner for celebrating the Passover—the popular passage known as "Four questions" (*Mah Nishtanah*), which must be addressed by a child to adults in the family, recites: "Why is this night different from all the other nights? … On all nights *we eat sitting upright or reclining, and on this night we all recline*?". Although the festival is certainly at least a millennium older, *Haggadah* was written in the present form around the third century CE, and this detail among others[6] reveals a clear Hellenistic influence.[7]

All these rules, as well as many others, constitute real *grammars of eating*, which are described in an abundant production of books about "manners", starting in Europe at least since Baldassarre Castiglione's *Libro del Cortegiano* (1528).

2.3 A General Semiotic Model for the Analysis of Alimentation

I will now introduce a sketch of a possible general model of analysis of the "alimentary process", based on the act of consumption. To this purpose, given the strong analogy, which many have noted, between food and language, I found it useful to consider the five steps of classical rhetoric (Cicero, *Rhetorica ad Herennium*; cf. Barthes 1970).

Rhetoric as an art of organising effectively and also making pleasant speeches go back to the oldest texts we have. Long before the roots of Western culture in the Bible and Homer, in Mesopotamian texts such as Gilgamesh, in the Egyptian and Sumerian inscriptions and in classical Chinese culture, there are particular ways of structuring a speech, figures of thought and expression that today we define according to the

---

[6] E.g. certain names such as *afiqomen*, from *epikomion*, the last piece of unleavened bread that is obligatory to eat at the end of the meal.

[7] Even if it celebrates the Jewish liberation and also cultural independence.

categories of rhetoric. But rhetorical theory, at least in the western tradition, was born later, and more precisely in Greece, at the same time as philosophy, history, theatre and democracy arose. It aims to identify those manipulations of the signifier that can make a speech *effective and pleasant*, not only in its actual oral expression, but also in its *preparation*. Unlike linguistic analysis, in other words, rhetoric involves the whole process—i.e. the empirical "generative path"—that leads to effective texts. Its problem, in semiotic terms, is *how to organize the expression of speech in order to obtain certain effects of meaning*. The analogy with alimentation is clear: food, like everything rotating around it, can be conceived as a "discourse", or "text", which in many cases aims to achieve specific effects of meaning. This is the reason why it is useful to study it from a semiotic perspective based on the five steps of classical rhetoric: *inventio* (the identification of the arguments—or foods); *dispositio* (their organisation and arrangement); *elocutio* (their shape and "style"); *actio* (their delivery—or consumption); *and memoria* (their remembrance/communication).

The phase of *inventio* would encompass:

1. *Social* collection and storage of edible matter:
   a Premises:
      i The *social definition* of edible matter and their syntagmatic compatibility (also including interdictions, which have been largely studied);
      ii The *social process of collecting food* (i.e. hunting, breeding, harvesting, fishing, gathering, chemical synthesising) and the principles regulating it: what foods can be collected? How frequently? What substances are considered more valuable?
   b The industrial or artisanal *transformation* and *preservation* (e.g. cutting, baking, freezing, smoking, salting, canning, adding sausages, preserving in oil) of edible matters;
2. *Individual* collection and conservation (for family life and/or single occasions) of food:
   a *Distribution* systems and technologies (i.e. markets, shops, large-scale distribution, online sale);
   b *Conservation* systems and technologies (i.e. refrigerators, larders).

The phase of *dispositio* concerns:

3. Planning:
   a Social decisions on *meal times* and *contents*:
      i The macro-syntagmatic level of seasons and days (i.e. holidays, celebrations, fasting);
      ii The medium-syntagmatic level of meals (i.e. breakfast, lunch, dinner);
      iii The micro-syntagmatic level inside meals (i.e. the succession or simultaneity of dishes and foods);
   b The *application* of these rules to actual food experiences.

4. Cooking:
   - a The choice of "dishes"/recipes (consequent to 2. and 3.b);
   - b The choice of ingredients:
     - i Main ingredients;
     - ii Solvents (e.g. milk, broth, water);
     - iii Seasonings and sauces (e.g. salt, fats such as oil, butter, or lard, spices such as herbs, pepper, *chile*).
   - c Cooking methods (e.g. "raw", rotten, roasted, boiled,[8] spit-roasted, baked, grilled, stewed);
   - d Cooking tools and technologies (i.e. pots, ovens, pans, microwaves, induction cookers), which are dependent on the tools and technologies of heat production (i.e. wood, coal, gas, or electricity);
   - e Secondary accessories (e.g. wooden spoons, potato peelers, colanders, mixers, blenders, knives of various types);
   - f Subsequent times, temperatures and treatments.[9]

The phase of *elocution* entails:

5. The arrangement of the space of consumption (i.e. places of the house, tables, seats) and the necessary equipment (i.e. tablecloths, cutlery, ornaments, seats);
6. Decisions about the diners (i.e. circumstances, invitations, gender ideologies, quality and hierarchy of the guests as related to social roles, age, etc., clothing).

Actio concerns:

7. Pre-eating peri-texts (e.g. prayers, toasts, cleaning practices, sacrifices);
8. Modes of service and consumption (e.g. shifts, times, choice and use of tools, conventions, ways of conviviality);
9. Peri-texts of consumption (e.g. music, conversations);
10. Post-eating peri-texts (e.g. prayers, cleaning practices, dances, entertainment).

Finally, *memoria* includes:

11. Pictures, letters, thanking remarks, counter-invitations.

This framework was conceptualised with regard to a somewhat formal dinner with guests in a modern western country. Of course, to better generalise this tool, other cultural contexts and other eating occasions should be considered. For example, we could consider larger and more formal cases (such as a public banquet) or simpler ones (from fast food to a couple's dinner in front of the TV, or at the restaurant). In any case, adaptation would be easy.

As a way of conclusion, I would like to suggest that this framework is not to be understood as an operational or technical list, but rather as a grid of possible

[8] Cf. (Lévi-Strauss 1964; Detienne and Vernant 1979).

[9] The whole process in the kitchen entails six main steps: preparing, cooking, serving, removing, washing, and putting away (see Frederick 1923).

foodsignifiers, that is to say, of *places of variability* or *points of attack of meaningful differences in the process of alimentation*—"differences that make a difference", therefore defining the sense of every single food experience. This approach gives alimentation a structural organisation that makes it not only a possible object of study for anthropologists, sociologists, scholars of customs, but also and above all for semioticians, because it highlights the social meaning not only of food, but of every step of the complex process conferring sense on it.

More specifically, I would like to suggest here three keywords for a further semiotic discussion of such a process: *performance*, *ritual*, and *expression*.

Performance is a more precise concept than "practice", which is widely used in semiotics. Although its earliest and basic meaning is "the accomplishment of an action", since the seventeenth century it has been associated with a range of theatrical practices and metaphors. A performance requires both the people (or even things) who perform and the people who witness the performance to exist. It is always *for* someone, even when the roles shift and the witness becomes the performer—or vice versa. Referring to a behaviour or an event as a performance therefore implies suggesting that they are in some way going *beyond* their simple functionality, and that they are consciously constructed for *appearing* in a certain way. Performance is *an action that consciously extends itself into the field of communication*, and is therefore designed to produce meaning both for those who realise it and for those who look at it. This is often the case with alimentation.

Performances often have *ritual* aspects. A ritual, in fact, is a ceremony or action performed *in a customary way in order to produce symbolic effects*. This concept is pertinent first and foremost in religious studies,[10] ethology, and sociology. But—I would suggest—also in semiotics, especially (even though not exclusively) in reference to food. Rituals have four main dimensions or tracts (Schultz and Lavenda 1987): (i) a repetitive social practice, which (ii) is different from the routines of daily life and (iii) follows some sort of schema (iv) encoded in a myth. Other characteristics of rituals often quoted are: formalism, traditionalism, invariance, rule governance, and performance (Bell 1997). This applies very well to alimentation. Formal meals, for instance, often take some appearance of rituals, as they are able to make transformations deemed significant by the societies within they take place. This applies symbolically, such as in religious ceremonies involving special forms of food, such as the Eucharist; but also on a social level, with meals used to celebrate alliances, friendships, weddings, seduction rituals, or even mourning. In short, the ritual aspect of meals allows their use as a "semiotic sanction" *par excellence* in social life.

Expression is a fascinating and important semiotic issue: there are processes of meaning-making that cannot be easily described by means of classical semiotic theories—neither those deriving from Saussure' or Greimas' models, nor those

[10] In religious studies, rituals and ceremonies are social events which are believed to have *the powerto effect meaningful transformations* (e.g. healing rituals). Rituals, not shared beliefs, provide the glue that holds together religious communities over a long period of time.

elaborated by Peirce—because their ability to signify does not depend on a narrative or sign structure, but rather from conformity or difference from social prototypes (Volli 2017, 2018). In alimentation and other complex social fields, such as fashion, architecture, or industrial design, we can find objects and performances that are certainly meaningful, but apparently not semiotically dense, as they seem to lack a non-trivial structure of signification or narration. Trying to find an expression/content structure or to identify specific actants in a certain attire or in a dish is often pointless. Yet, they are able to produce sense. People are keen to appear in these performances and to use these objects *in the right way*, because what is at stake is their social definition. These practices and objects do not correspond, if not in part, to the common definition of communication as *transport of contents to precise recipients*, nor to the above-mentioned semiotic structures and tools of analysis. But—following Umberto Eco's definition of the limits of communication and semiosis—we can use them in order to *lie*. Performances and rituals, in particular those related to alimentation, work according to a *principle of expression*, that is to say, to their *appearance* in a *dialectical relationship* with *recognised rules and social models*. Their meanings emerge from their *conformity to* or *distance from* such rules and models, which identify specific *positions*, and hence social subjects, describing them as "hegemonic", "resistant", "experimental", "fashion-addicted", "insiders", "outsiders", "aristocratic", "bourgeois", and so on.

More work therefore needs to be done, on an empirical but also on a theoretical level, in order for semiotics to be able to adequately describe these *expressive* forms of production of meaning. This seems to urge us to extend the field of food studies from food itself to the broader domain of alimentation, as well as to elevate the semiotic analysis of daily-life phenomena from the level of practices to that of performances, and from a purely textual methodology to expressive analysis.

References

Barthes, Roland. 1970. L'ancienne *rhétorique*. Aide-mémoire. *Communications* 16: 172–229.

Bell, Catherine. 1997. *Ritual: Perspectives and Dimensions*. New York: Oxford University Press.

Casey, Wilson. 2009. *Firsts: Origins of Everyday Things that Changed the World*. London: Penguin Books Ltd.

Detienne, Marcel, and Jean-Pierre. Vernant. 1979. *La cuisine du sacrifice en pays grec*. Paris: Gallimard.

Elias, Norbert. 1939. *Über den Prozeß der Zivilisation. Soziogenetische und psychogenetische Untersuchungen. Erster Band. Wandlungen des Verhaltens in den weltlichen Oberschichten des Abendlandes*. Basel: Verlag Haus zum Falken. English edition: Elias, Norbert. 1969. *The Civilizing Process, Vol. I. The History of Manners* (trans: Jephcott, Edmund). Oxford: Blackwell.

Frederick, Christine. 1923. *Household Engineering: Scientific Management in the Home*. Chicago: American School of Home Economics. https://archive.org/details/householdengine00fredrich/page/n5/mode/2up. Accessed 3 June 2020.

Lévi-Strauss, Claude. 1964. *Mythologiques*, t. I : *Le Cru et le Cuit*. Paris: Plon.

Lévi-Strauss, Claude. 1968. *Mythologiques*, t. III : *L'Origine des manières de table*. Paris: Plon.

Marx, Karl. 1858. *Grundrisse: Foundations of the Critique of Political Economy* [Engl. Transl. 1993]. London: Penguin.
Post, Emily. 1950 *Etiquette: the blue book of social usage.* New ed. completely rev. [First published. 1922. *Etiquette in society, in business, in politics and at home*]. New York: Funk & Wagnalls Co.
Ranhofer, Charles. 1894. *The Epicurean*, New York: Ranhofer. Now online: https://archive.org/details/epicureancomplet00ranh. Accessed 3 June 2020.
Rebora, Giovanni. 2013. *Culture of the Fork: A Brief History of Everyday Food and Haute Cuisine in Europe.* New York: Columbia University Press.
Schultz, Emily A., and Robert H. Lavenda. 1987. *Cultural Anthropology: A Perspective on the Human Condition.* London: Oxford University Press.
Segre, Anna. 2001. *La Pasqua ebraica.* Torino: Silvio Zamorani Editore.
Stano, Simona. 2015. *Eating the Other: Translations of the Culinary Code.* Newcastle-upon-Tyne: Cambridge Scholars Publishing.
Stano, Simona. 2018. *I sensi del cibo.* Roma: Aracne.
Visser, Margaret. 1992. *The Rituals of Dinner*. London: Penguin.
Volli, Ugo. 2014. Pedagogy and enunciation in religious intrinsically coded acts. The case of the Passover Seder. In *Proceedings of World Congress of IASS-AIS*, ed. Kristian Bankov. Sofia: NBU Press. https://doi.org/10.24308/iass-2014-091.
Volli, Ugo. 2016. Taste and meaning. *Semiotica* 211: 231–246.
Volli, Ugo. 2017. Sense and Marking. In *Semiotics and its Masters*, ed. Kristian Bankov and Paul Cobley. 223–235. Boston-Berlin: De Gruyter.
Volli, Ugo. 2018. Segni anomali. In *Semiotica generale – semiotica specifica*, ed. Artur Gałkowski and Tamara Roszak, 38–48. Lodz: Uniwersytet Łódzki.
Volli, Ugo. 2019. Il problema semiotico e filosofico del gusto. *E/C* 26: 107–116.
Wilson, Bee. 2012. *Consider the Fork: A History of How We Cook and Eat.* New York: Basic.

Ugo Volli is Honorary Professor of "Semiotics of the Text" and "Philosophy of Communication" at the University of Turin. He taught in a number of international universities, including Bologna, IULM, Brown, Haifa, New York University. He was awarded an honorary doctoral degree from the New Bulgarian University. He is the editor in chief of *Lexia—Journal of semiotics* and member of scientific boards of many international journals. His last books are *Periferie del senso* (Aracne 2016) and *Il resto è commento* (Belforte 2019).

Chapter 3
On the Face of Food

Massimo Leone

> "I do wish we could chat longer, but I'm having an old friend for dinner".
> Hannibal Lecter in *The Silence of the Lambs* (Jonathan Demme, 1991)

Abstract Face and food are both essential interfaces of social life. Their intersections in history and cultures are still underexplored. Humans see faces in food, as in the cognitive phenomenon of pareidolia; they turn food into the face of deities, as in Pastafarianism; they create artworks in which food compose faces, like in Arcimboldo; the face of non-human animals turned into food is often concealed, whereas other kinds of food, like the Japanese "character bento" and "chigiri-pan", are anthropomorphized through the attribution of a face. Disquietingly, certain drugs, like the so-called "bath salts", seem to urge users to cannibalize the face of other people. Face turned into food, food turned into face, face removed from food, face instilled in food: what happens, from the semiotic point of view, when food is visually given a face and what, on the opposite, when this face is hidden?

Keywords Face · Food · Semiotics · Exhibition · Occultation · Vegetarianism

---

A first version of this article was presented as plenary lecture in the symposium "Food for Thought: Nourishment, Culture, Meaning", New York University, 14–15 October 2019. I thank the two organizers, Prof. Amy Bentley and Prof. Dr. Simona Stano, for the wonderful opportunity. This project has received funding from the European Research Council (ERC) under the European Union's Horizon 2020 research and innovation programme (grant agreement No 819649 - FACETS).

M. Leone (✉)
Department of Chinese Language and Literature, University of Shanghai, Shanghai, China
e-mail: massimo.leone@unito.it

Department of Philosophy and Education Sciences, University of Turin, Turin, Italy

S. Stano and A. Bentley (eds.), *Food for Thought*, Numanities - Arts and Humanities in Progress 19, https://doi.org/10.1007/978-3-030-81115-0_3

3.1 Sweet Simulacra

Hopefully it will not read inappropriate to begin this article with two personal anecdotes, one from my childhood in Southern Italy and the other one from a more recent past. For Christmas and for Easter, people of Lecce, in Apulia, "the heel of the Italian boot", keep the tradition of eating sweets made out of *pasta di mandorla*, literally "almond paste", which is similar to marzipan but contains more ground almonds and less sugar.[1] For Christmas, pastry shops in the city sell cakes in the shape of fish, whereas for Easter, cakes are sold in the shape of lambs. Both come in different dimensions and some of them are particularly lavish (Figs. 3.1 and 3.2).

A former assistant of my father used to give one to my family as a gift for Christmas and then again for Easter: huge fish and gigantic lambs would then emerge from cardboard trays wrapped in immaculate silky paper and elaborately tied ribbons, their scales and mouths and eyes and tails or, for Easter, their curly fleece and little ears, recreated through delicate strokes of pale chocolate, their profile neatly sculpted into the sweet paste, often filled with pear jam or with *faldacchiera*, a concoction of slowly cooked yoke, pear jam, and chunks of dark chocolate. The most skillful cooks, usually the grannies, would prepare these sweets at home, using molds with the appropriate shape and form, but the best *pasta di mandorla* in Lecce, my hometown, is still made and sold by "le Monache", as locals simply call them, that is, the cloistered Benedictine nuns of the beautiful baroque convent of Saint John the Evangelist, nuns who, like many cloistered devotees in the Catholic world, dedicate themselves to the production of sweets.[2]

Their fish and lambs are, indeed, exquisite; furthermore, buying them through the wheels of the nunnery, that is, the wooden revolving windows through which, for centuries, the pious women have kept secretive and selective contact with the

[1]On the food of Lecce, see Foscarini (1987); Patience Gray, the food writer and author of *Honey From a Weed* (1986), briefly wrote about these cakes (1977: 30); see also Tarantino and Terziani (2010).

[2]For a history of the Convent of the Benedictine nuns of Lecce, see De Meo et al. (2006). Mary Taylor Simeti, the food writer and historian of Sicilian food, wrote a book that recollects the memories and recipes of Maria Grammatico, one of the nuns of the cloistered Istituto San Carlo, Erice, Sicily, also devoted to the production of confections (Grammatico and Simeti 1994). The book, *Bitter Almonds: Recollections and Recipes from a Sicilian Girlhood*, narrates the story of a woman, Maria Grammatico, sent as a child to a cloistered orphanage in Erice, on the western coast of Sicily, where she learned to prepare the handcrafted pastries, especially marzipan confections, that were sold to customers outside the convent walls. As Mary Taylor Simeti points it out in the introduction to her book, the tradition of nuns preparing such sweets might be very old: there is historical evidence that, at the end of the sixteenth century, nuns of the Diocese of Mazara del Vallo, in Sicily, were prohibited to make cassata, the typical Sicilian cake with ricotta and almond paste, during Holy Week, lest the preparations distract them from prayers. Systematic production of confections in cloistered convents, though, began in the 1860s after the newly formed Italian State confiscated the Catholic Church's properties, pushing convents to find revenues in alternative ways, among them the making and selling of almond cakes. The second chapter of Simeti (1986) is devoted to the story of these delicacies.

Fig. 3.1 Almond-paste cake in the shape of fish sold in Lecce for Christmas (From https://en.wikipedia.org/wiki/File:Pesce_di_pasta_di_mandorle.jpg. Last accessed 22 November 2019)

Fig. 3.2 Almond-paste cake in the shape of lamb sold in Lecce for Easter (From https://commons.wikimedia.org/wiki/File:Agnello_pasquale_pasta_mandorle_salentino.jpg. Last accessed 22 November 2019)

world (wheels that were once used also in order to entrust abandoned orphans to the care of the nuns), adds a ceremonial, mysterious, and almost sacred touch to the festive pleasure of purchasing these sweets: from the semi-dark hole in the thick baroque walls of the convent, a shadowy hand emerges, usually old and pale, holding a simple package tied with an ordinary cord.[3] The minimalist wrapping, though, enshrines the best, and also the most expensive, *pasta di mandorla* of Lecce.

The sacredness of the gesture of buying these almond fish and lambs indeed is not diminished by the commercial exchange. As a matter of fact, these cakes are meant to be deeply connected with the two most important Christian festivities, at the beginning and at the end of Jesus's life; their shapes, moreover, are evidently related to Christian iconography: with the fish graffiti that, according to legendary tradition, the still persecuted Christians would draw in the Roman catacombs and in other secret places so as to mark their cryptic religious affiliation (a secretive marking that, by the way, was also adopted by seventeenth-century persecuted Christians in Japan)[4]; but also with the lamb,[5] that is, the sacrificial animal *par excellence* in the Abrahamic religions, an ancestral and probably even pre-monotheistic tradition according to which the fidelity of the religious community to its transcendence is to be periodically reaffirmed through the ritual killing of a member of the group, usually the most defenseless one, so as to symbolically fortify, through the sacrifice of the designated individual, the sacred cohesion of the group.[6]

---

[3]Carlo Levi, famously author of the novel *Cristo si è fermato a Eboli* [Christ Stopped at Eboli] (1945), describes buying almond-paste cakes from a cloistered convent of Erice in the book *Le parole sono pietre: Tre giornate in Sicilia* [translated as "Words are Stones: Impressions of Sicily"] (Levi 1955): "We wanted to taste the famous cakes of almond paste and the mustazzoli made by the nuns of a cloistered convent. We entered the atrium and expressed our wishes to a sort of shadow behind a double grate and shortly, without any accompanying words, the pastries appeared on the wheel, tender flowers of green and pink and violet and azure, and we left our money in their place" (Engl. trans. 1958: 165).

[4]See Stroumsa (1992) and Rasimus (2012); on Japan, see Leone (2018c).

[5]See Nikolasch (1965), Skaggs and Doyle (2009), De Lang and Marijke (2017), and Benarroch (2019).

[6]Furthermore, as pointed out by Tarantino and Terziani (2010): "In the Christian tradition the almond is a symbol for the soul, and its oval shape encircles holy figures in medieval imagery. When nuns in Sicilian convents use finely ground almonds for their feats of confectionery, we can experience the incarnation of the spiritual raised to the nth degree" (49).

3.2 Sweet Sacrifices

Only the group that is able to sacrifice one of its own members will be able to secure the favor of the deity.[7] Yet, in monotheisms as well as in other religious traditions, the mechanical force of the human sacrifice is replaced by the symbolical force of the animal sacrifice. The animal is offered to the deity because its spiritual appetite is molded after that of its human devotees, but also because, through the animal, a replacement takes place.[8] The animal is, indeed, a simulacrum. In many ancient cultures, and to a lesser extent also in the present-day societies, the non-human animal is not given the same dignity of a human. It is used as food, but even as food, its primary purpose is to allow the displacement of tension and conflict within the community: eating non-human animals, but even more sacrificing non-human animals, avoids the tragic embarrassment of seeing a fellow human being, a member of one's own community, as the next meal, or the next sacrificial victim, and avoids also the risky endeavor of finding material and symbolical food through war. In the sublimation of the human sacrifice, as well as in the sublimation of cannibalism, the human community creates, in the topology of its spiritual space, a meta-level, in which both the starving of humans and that of the deity can be appeased without resorting to human bloodshed. Killing non-human animals might indeed be the most fundamental traditional instrument of social cohesion.

When families of Lecce gather together for the lavish festive meals of Christmas and Easter, they eat and drink through extravagant banquets, which culminate in the apical moment when someone, usually the father, unwraps the white package, discloses the wonderful sweet animal inside, either a perfectly carved fish or a perfectly carved lamb, and presents it to the family: at this moment, exclamations of marvel ensue, if the simulacrum appears as well shaped and colored, but also manifestations of hilarity burst, when some home-made almond fish or lambs manifest the countenance of bulky whales or bulls, due to the poor artistic skills of the local granny.[9]

The shape is, indeed, important. In traditional southern Italian families, as in many traditional families across the religions of the book around the Mediterranean, a true sacrifice would take place for Easter. Christian housewives in Lecce would try to secure, often weeks in advance, the youngest suckling lamb from the local butcher; some would even choose it directly at the farm, and then wait for Easter so as to bake it and offer it to the family as the main dish of the ritual meal. That killing, cooking, and eating of the little animal was in line with a millenary tradition of sacrifice, replacing the innocent child with the innocent lamb, and quenching,

---

[7]Literature on sacrifice is extensive. Among the most recent and relevant contributions, see Murray (2016), Alison and Palaver (2017), Pesthy-Simon (2017), Foraboschi (2018), and Terrin (2019).

[8]See García Fernández et al. (2015) and Van Straten and Folkert (2016).

[9]The hilarious Facebook page "Agnelli di pasta di mandorle brutti" collects images of these missshapen sweets.

through the killing, the material and symbolical thirst of the deity, which is nothing but a counterpart of the symbolical thirst of the community.

3.3 A Semiotics of Sacrifice

As the meta-level of the animal sacrifice would sublimate, thus, the basic level of the human ritual killing and eating, so the communal eating of the lamb-shaped almond cake would take the sacrifice to a third meta-level. Whereas anthropologists study these superimposing dimensions through gathering evidence about their ethnological and social instantiations,[10] the semiotician is rather interested in the language of sacrifice: what signs are necessary, so that the sacrifice maintains its identity, even in a sublimated sphere?[11] At each meta-level of sublimation, indeed, the semiotic setting of the sacrifice loses some properties while acquiring some other features. The sacrificed animal shares with the sacrificed human two essential characteristics, which are life and motility. The two are connected: animals use their motility exactly to prolong their lives, seeking either to chase food or to escape being chased as food[12] Life, in turn, can be defined in relation to motility: everything that, in the universe, prolongs its existence by purposefully moving through space, is alive.

Sacrificed animals, like sacrificed humans, add to the symbolic efficacy of the ritual exactly because they would rather escape, and exert their motility so as to run away from the priest, or from the butcher, so prolonging their, the animals', lives. The energy of the sacrifice is due to this resistance, an energy that would be lessened, or even disappear, should the animal victim be replaced with an inert one. Religions are rarely interested in the sacrifice of plants. Vegetables and fruits are offered to the deities in many religions but are not properly sacrificed to them. That is because vegetables too purposefully move in the environment so as to prolong their lives, for instance through stretching their roots in search of water or their leaves in search of light; yet, the degree of this motility is lesser than that of animals not only quantitatively, but also qualitatively: vegetables do not make good sacrificial victims not only because they can move in space but not through space, and could not, as a consequence, seek to escape from the sacrificial knife; indeed, the qualitative nature of their motility is also different: they purposefully move in space so as to seek more nutrients, but they do not do so intentionally.[13]

Indeed, the more capable of intentionality a victim is, the more powerful the sacrifice. Christianity has stretched this equation so far as to instituting its entire theological identity around the figure of a self-sacrificing god: a plant cannot escape

[10] Among the most recent contributions, see (Kitts 2018).

[11] For a sketch of semiotic typology, see Hastings (2003) and Janowitz (2011).

[12] For a theoretical introduction to a biosemiotics of motility, see Leone (2012).

[13] For a semiotic perspective on intentionality, see Leone and Zhang (2017).

ritual killing; a non-human animal can; a human can even, contrarily to most animals, voluntarily decide to die in sacrifice, hence the symbolical powerfulness of martyrs for the shaping of a spiritual community[14]; but nothing can beat the paradoxical image of an omnipotent god that lets his own son offer himself in sacrifice: that is the ultimate sacrificial scenario, because it combines the highest potentiality of resistance to death with the humblest acceptance of it. It is a sacrifice that epitomizes the Christian idea of community.

3.4 Sacrifice and Sacred Face

The lamb-shaped almond cakes of Lecce do not emanate the same symbolical aura; they cannot move, suffer, or try to escape, and offer themselves placidly to the knife of the father. Yet, if they are just a sweet but inert ersatz of living lambs they are, nevertheless, symbolically more than simple objects. Their being shaped in the effigy of lambs, indeed, turns them into strange creatures, into artifacts that, while being deprived of purposefulness and motility, like plants, and even more of intentionality and motility, like animals, and even beyond that, of language and motility, like humans and humanoid deities, lamb-shaped almond cakes are, nevertheless, endowed with agency.[15] The sacrificial potential of such cakes is lesser than that of the animals that they, the cakes, represent, which is lesser than that of the humans that they, the animals, replace, which is lesser than that of the gods that they, the humans, imagine. Yet, these cakes are better sacrificial victims than plants. The former, by virtue of their shape and iconic resemblance to the animal victims they stand for, can exert a symbolical action that is precluded to the latter. The main hypothesis of this chapter is that such symbolical action, and the sacrificial potential that it underpins, is strictly related to the iconic emergence of a face, which is in turn complexly related with the iconic emergence of a gaze, which is in turn also related with the emergence of some eyes.

Again, some anecdotal evidence will lead to the formulation of such hypothesis. When I was a child, my father would make the first cut to the almond-made fish, or to the almond-made lamb, and this initial cut would always be inflicted at the central part of the animal body, right in the belly. My mother would jokingly say that my father, a mighty sugar-eater, would do so in order to get the thicker slices, replete with pear jam, but, in reality, nobody in Lecce would cut these cakes starting from the head. Once that I, still a child, jokingly decapitated one of these sugary lambs, I was scolded by my mother. That was not supposed to be done, I was told.[16] There were even superstitions, if I remember well, related to eliminating the head of

[14]See Leone (2018b).

[15]See the classic Gell (1998).

[16]In Christianity, beheading is inflicted on many martyrs, including Saint John the Baptist, but not to Jesus; cfr Baert and Rochmes (2017).

one of these sweet effigies too soon. On the contrary, one was supposed to keep cutting from the middle towards the head and the tail, until these cakes were actually turned into monsters composed of head and tail only, like those that swarm in the paintings of Bosch. Indeed, in Italian as in English, people say of something that "non ha né capo né coda", "neither head nor tail", to point at its indecipherability. More precisely, this way of cutting the cake was instrumental to preserve its gestalt and, most importantly, the face of the effigy.

3.5 En-visaging and De-facing

A general research field that an ERC project of mine currently explores is the dialectics between, on one hand, visual processes of *en-visaging*, as I call the semiotic operation of attributing a visage to something or someone, and *de-facing*, as I call the opposite semiotic operation of eliminating something's or someone's face. In my mind, this dialectic is related to the one between anthropological practices of humanization and de-humanization. This both visual and socio-cultural dynamic is central, I believe, also in the complex symbolic domain that connects food, sacrifice, and the community.

What are, then, the relations between face and food? First, a face can hardly become food, for two reasons mainly. On the one hand, most cultures and their respective languages do not conceive of edible vegetables and non-human animals as being endowed with a face. A face is attributed to vegetables only in specific circumstances that do not normally include eating them; carrots, zucchinis, eggplants are not usually "seen" by cultures as having faces, although, as we shall see, they can be given one under special conditions, as it famously happens with pumpkins for Halloween; as regards non-human animals, most cultures distinguish, also in language, between the face of a human and the equivalent of a non-human animal, for the designation of which usually another word is used, like "muzzle" in English or "muso" in Italian. A muzzle is a kind of a face but it is also less than it: it is a face stripped of the phenomenological, aesthetic, and especially ethical aura that characterizes faces; that is why, when we want to ethically "promote" a non-human animal, we start consider its muzzle as a face, as it commonly occurs with pets. The converse is also true: when we want to demote a human being, we start consider her or his face as a muzzle, as it is the case of several idiomatic expressions in many languages, such as, in Italian, "brutto muso", literally, "ugly muzzle". In Italian it can be said of animals, and even of human beings, although usually of women, that they have a "bel musetto", literally, a "nice muzzle", but a "brutto muso" is normally attributed to humans with derogatory intents. In any case, humanized animals usually receive a face, whereas dehumanized human beings usually receive a muzzle.

In both circumstances, however, edibility is not an option. Neither a face nor, to a lesser and different extent, a muzzle can be eaten, because what they designate is usually not simply a part of the body but the phenomenology of it in social interaction, resulting from the exposition and arrangement of this body part, that is, the frontal side of the head, to others, or to a mirror, or even to an imaginary other (as it occurs when we modify our face while talking to ourselves). Hence, the face cannot be eaten not only because it is human, for the muzzle cannot be properly eaten either. Neither of them can be eaten because they are intrinsically alive: a face, indeed, emerges from the conjunction of a head, with all its component, and life. Thus, it is actually the head of a person or of a non-human animal that can be eaten, or even the components that cultures and languages single out in such heads, like mouths, ears, eyes, chins, cheeks, or even inside of heads, like tongues or brains; the difference between eating a head and eating a face is very similar, although not identical, to the difference between eating an eye and eating a gaze: some food cultures turn eyes into food, but gazes can be eaten only metaphorically and, actually, they more often take an active part in the metaphor; we can say of someone that she or he eats someone else with her or his gaze, but rarely the opposite is said: nobody's gaze is usually eaten by nobody, for the simple reason that eating something *ipso facto* implies the end of its motility, intentionality, and life, so that eating something that is alive expresses a paradoxical and, as a consequence, extreme possibility in food cultures.

Indeed, the more cognitively, emotionally, and pragmatically complex a living being is, the less socially acceptable it is to turn it into food while it is alive. Whereas fruits are eaten when they are technically still alive, although with the kind of life that can be imputed to vegetables, only in rare circumstances are animals, including non-human animals, eaten alive. In cannibalism, the human body is usually already dead when it is turned into food, also for a living human would probable resist its being turned into an edible matter; incidentally, that also introduces a difference between normalized cannibalism and abnormal cannibalism: Hannibal the cannibal, as well as vampires, disconcert not only because they feed on human flesh and blood, but also because in many cases this happens when the victim is still alive, moving, and suffering.

Disquietingly but interestingly, moreover, certain drugs, like the so-called "bath salts", seem to urge users to cannibalize the face of other people. More commonly, some non-human animals are also sometimes eaten alive, for instance worms in certain kinds of ultra-fermented cheese, octopuses in some particularly extreme Japanese recipes, or molluscs; in all these cases, though, the faces of what we eat are not seen or they are barely so. With the exclusion of these abnormal exceptions, though, what cultures eat is not normally faces as actuality but faces as potentiality, dead faces that cannot be used as faces anymore and in which, moreover, the capacity of acting as faces is downplayed or eliminated.

3.6 Heads

What many food cultures eat, indeed, is heads, not faces. The face disappears not only when the head is stripped of its life and, therefore, its expressive potential, but also as a consequence of cooking. A face is supported by a skeletal structure but its own substance must be fleshy and malleable, exactly for the purpose of moulding itself into many expressions. Although it is exactly this soft, expressive substance that cultures sometimes turn into food, they systematically transform it before doing so. Cultures eat skin, and sometimes they even consume it without separating it from its flesh, but that is rarely done with heads. Many traditional societies eat mutton heads, for instance, yet when we see these heads, when they are cooked or even when they are on display in a butchery, ready to be cooked, what we see is usually a bloody conjunction of skull, tendons, muscles, and facial organs that hardly let the idea of a proper face emerge.

There is a complex but systematic relation between the transition of societies into modernity and the inclination to turn heads into food. In many modern societies, eaters are displeased or even disgusted at the sight of heads of non-human animals turned into raw food (like mutton heads); they are disquieted at the sight of animals turned into food and served with their head visible; and, in most cases, they would never accept eating the head or the facial parts of a non-human animal. On the one hand, this rejection of the head as food shows some latent hypocrisy. Once I was in a Piedmontese restaurant with Umberto Eco. In traditional Piedmont restaurants, as in many other parts of world, the tongue is still eaten, for instance in the famous "bollito misto". As one of the diners, a lady, would protest that she would never eat a tongue because it had previously been in the mouth of an animal, Eco yelled at the waiter: "Waiter, that is disgusting, we cannot eat something that was in the mouth of an animal; please, bring us eggs instead!".

I must say, however, that having spent much of my pre-vegan time in Iranian contexts, where the tongue is a delicacy and is often prepared with many different recipes, what mostly puzzled me while eating it was not its previous location, but its previous function: that tongue had been not only part of a mouth but also part of a face, meaning that it was instrumental to create its expression, and it was now turned into food and chewed into my own mouth by my own tongue. This kind of considerations are recurrent in present-day reactions to the prospect of eating what many pre-modern food cultures would eat without too much perplexity: not only heads, then, but also tongues, chins, cheeks, ears, whole muzzles, and even eyes.

3.7 A Patterned Geography of Head-Eating

A complex but patterned geography of face-eating can then be drawn, depending on how societies relate to the idea and the practice of feeding on head- and facial parts. Such map should be not only spatial but also temporal, and take into account differences in terms of class and gender.

Khash (Armenian: խաշAzerbaijani: xaş; Georgian: ხაშიKhashi), *pacha* (Persian: پاچه;Albanian: paçe; Arabic: باجة;Bosnian: pače; Bulgarian: пача;Greek: πατσάς),*kalle-pache* (Persian: کله‌پاچه;Turkish: kelle paça), *kakaj šürpi* (Chuvash: какай шӳрпи)or *serûpê* (Sorani Kurdish: سه روپێ)is a dish of boiled cow or sheep parts, which often includes the head. It is a traditional dish in Afghanistan, Albania, Armenia, Azerbaijan, Bosnia and Herzegovina, Bulgaria, Georgia, Greece, Iran, Iraq, Mongolia, and Turkey, but with different popularity in the present-day societies of these countries.

Whereas the heads of *kalle-pache* are still commonly on display in a busy street of present-day Iran, Italians would now mostly be squeamish about the public offer of animal heads. Compare, for instance, the way in which a popular Italian website of recipes, *Cuoca a tempo perso*, present two lamb heads in one of its pages.[17] The juxtaposition of the smiling face of the cook and author, touching her cheeks in a relaxed posture, and the decorticated heads of two little lambs in the recipe picture, should indeed be shocking, but the contrast is somehow anesthetized by the Instagram-like composition of the ingredients: the two heads harmonically arranged on a perfectly white chopping board, with no stains of blood whatsoever; the gentle touch given by the display of innocent herbs, both in a traditional basket and beside the heads; even the Italian name of these body-parts, which in the title of the recipe are not called "teste", that is, "heads", but "testine", that is, with a diminutive form, "little heads", as though to beautify the beheading of the two kids. Despite this edulcorating presentation, though, the writer feels urged to explain:

> The baked little head of lamb is a classic of Italian regional food and among the recipes that are a little hard: old flavors, by now forgotten, which we rarely reproduce at home. A little bit because of the difficulty of finding the ingredients, but above all [we can say it], because we are squeamish about eating certain kinds of food. This special ingredient, though, is a stronghold of poor food, where nothing is thrown away and little is needed to turn a very poor kind of food into a King. Offer it to the table the day of Easter together with the rest of the lamb and certainly you will find some aficionados.[18]

---

[17]http://www.cuocaatempoperso.com/2018/03/testina-di-agnello-al-forno-con-erbe.html (last accessed 22 November 2019).

[18]"La testina di Agnello al forno e' [sic] un classico di tutte le cucine regionali e le cucine un pò [sic] hard: vecchi sapori oramai dimenticati che raramente rifacciamo nelle nostre case. Un po' forse diciamo per la reperibilità di prodotti ma soprattutto perché (si può dire) ci fa senso mangiare certi alimenti.

Pero' [sic] da sempre questi alimenti cosi [sic] particolari sono stati un caposaldo delle cucine povere dove non si butta nulla e dove basta poco per far diventare Re un alimento molto povero.

That, however, might not be true in present-day Italy anymore. Lamb eating is being increasingly stigmatized. This traditional food is more and more avoided even during Easter banquets. Offering lamb heads to family and friends over Easter lunch might be severely frowned upon. Indeed, more and more, the West wants to conceal the face of animals while eating their meat, especially if such animals, as in the case of lamb, are unanimously considered as "cute". An aristocracy of beauty, indeed, reigns also among non-human animals. Moreover, whereas the indexical link between meat and the animal bodies whence it stems is concealed through packaging,[19] the packaging of heads and faces as meat is increasingly rare, and judged as a monstrosity in most present-day post-modern societies.

3.8 Conclusions

Turned into an instrument of cultural semiotics, though, this map should lead to more general questions than the simple distribution of taste and distaste across history and geography. The fact that a society stops eating the head and facial parts of non-human animals, especially of mammals but also of birds and fish, reveals a deeper trend, which is at the same time aesthetic and ethical. On the one hand, societies tolerate less and less being reminded that what they are eating used to be alive and have a face; on the other hand, this hypocritical removal of the face from food is complexly linked with the decline of the idea of food as sacrificial matter, as sacred matter, and as the centre of a spiritual and ritual practice. With modernity, human cultures seem to repress more and more the idea that, when eating non-human animals, they are actually using them as sacrificial ersatz of human beings. Eating animals is what allow humans not to feed on each other, yet the unconscious cultural inclination to strip animals of everything that they have in common with humans, and therefore primarily of their faces, is instrumental to this practice of substitution and repression. The part of humanity that is in animals is eliminated so that the part of animality that is in humans may be preserved. Vegetarians and even vegans though are not immune from this symbolic circle, which is the cultural expression of the biological necessity of feeding life with life: outside of the sphere of language, this necessity is just part of the fabric of nature, yet within language, and therefore within cultures, the natural need of feeding life with life immediately turns into the cultural tragedy of having to feed life with death. Since our birth, we survive thanks to the death of other previously living beings, but modern individualism cannot cope with this sense of dependence, which in pre-modern and especially spiritual cultures would give rise to a whole series of rituals and practices meant to appease the mourning of food, for instance through

Mettetela in tavola il giorno di Pasqua insieme al resto dell'agnello e troverete sicuramente qualche appassionato".

[19]For a thorough examination of the semiotics of meat and fish packaging, see Leone (2018a).

praying before, while, and even after producing or consuming food, or through ritual fasting.[20]

The post-modern society celebrates food more and more; it turns the production of food into a feast, its consumption into a joy, its refinement into a cult. Everything that reminds eaters of the part of death that inevitably food entails is repressed, removed, concealed. The face of animals disappears. It reappears only in simulacrum, as a fiction, to nervously reassure eaters that they are feeding on the image of a face, and not on a face itself. The post-modern society celebrates food more and more, and it should continue to do so. After all, eating is what allows us to be, both in nature and in culture. Yet, one is left wondering whether societies should also, to a certain extent, mourn food, and while accepting the impossibility of eliminating the death of food from the food of life, acknowledging it. Acknowledge the suffering that our food is made of, that our bodies are made of, that even our thoughts are made of. That would perhaps partly diminish the arrogant modern pleasure of treating food as simple matter, but would remind eaters that what they eat in food is not only materiality, but also spirituality, the spirituality of what had to stop being so that we might continue being, of life that ended so that life might continue.

If that is not a definition of what a sacrifice is, then what is?

References

Alison, James, and Wolfgang Palaver, eds. 2017. *The Palgrave Handbook of Mimetic Theory and Religion*. New York, NY: Palgrave Macmillan.

Baert, Barbara, and Sophia Rochmes, eds. 2017. *Decapitation and Sacrifice: Saint John's Head in Interdisciplinary Perspectives: Text, Object, Medium* (Art & Religion: 6). Leuven, Belgium: Peeters.

Benarroch, Jonatan M. 2019. 'Christum qui est Hædus Iudaeis, Agnus Nobis': A Medieval Kabbalistic Response to the Patristic Exegesis on Exodus 23:19. *The Journal of Religion* 01 July: 263–287.

De Lang, Marijke H. 2017. John 1.29, 36: The Meaning of Ἀμνὸς τοῦ Θεοῦ and John's Soteriology. *The Bible Translator* 68(2): 148–63.

De Meo, Antonio, Rosellina D'Arpe, and Roberta Dell'Anna, eds. 2006. *Carte per la vita e carte per la storia: Le Benedettine in Lecce tra Medioevo ed età barocca*. Lecce: Milella.

Foraboschi, Daniele. 2018. *Violenze antiche*, ed. Silvia Bussi. Wiesbaden: Harrassowitz Verlag.

Foscarini, Antonio Edoardo, ed. 1987. *La cucina rusciara. Curiosità, tradizioni e ricette della cucina leccese*. Lecce: Conte Editore.

García Fernández, Francisco José, Fernando Lozano Gómez, and Álvaro Pereira Delgado, eds. 2015. *El alimento de los dioses: Sacrificio y consumo de alimentos en las religiones antiguas* (SPAL Monografías: 20). Sevilla: Universidad de Sevilla.

Gell, Alfred. 1998. *Art and Agency: An Anthropological Theory*. Oxford, UK and New York, NY: Clarendon Press.

Grammatico, Maria, and Mary Taylor Simeti. 1994. *Bitter Almonds: Recollections and Recipes from a Sicilian Girlhood*. New York: W. Morrow.

Gray, Patience. 1977. Little Fishes of the Nativity, *Harpers and Queen*, December: 30.

[20]See Leone (2013).

Hastings, Adi. 2003. From Ritual to Grammar: Sacrifice, Homology, Metalanguage. *Language and Communication* 23(3): 275–285.
Janowitz, Naomi. 2011. Inventing the Scapegoat: Theories of Sacrifice and Ritual. *Journal of Ritual Studies* 25 (1): 15–24.
Kitts, Margo. 2018. *Martyrdom, Self-Sacrifice, and Self-Immolation: Religious Perspectives on Suicide*. Oxford, UK: Oxford University Press.
Leone, Massimo. 2012. Motility, Potentiality, and Infinity: A Semiotic Hypothesis on Nature and Religion. *Biosemiotics* 5: 369–389.
Leone, Massimo. 2013. Digiunare, istruzioni per l'uso: La mistica dell'inedia nel Giainismo. *E/C* 14: 47–58.
Leone, Massimo. 2018. Food, Meaning, and the Law: Confessions of a Vegan Semiotician. *International Journal for the Semiotics of Law* 31 (3): 637–658.
Leone, Massimo. 2018. Il martirio interiore: Segni e testi del sacrificio di sé nel primo Cattolicesimo moderno. *Lexia* 31–32: 103–121.
Leone, Massimo. 2018c. Il sacro nascosto: studio semiotico sui Kakure Kirishitan [隠れキリシタン]. *E/C*, online journal of the Italian Association for Semiotic Studies, 23 December 2018: 1–13. http://www.ec-aiss.it/index_d.php?recordID=899. Accessed 21 June 2020.
Leone, Massimo, and Jiang Zhang, eds. 2017. *Intenzionalità / Intentionality*, special issue of *Lexia*, 29-30. Rome: Aracne.
Levi, Carlo. 1955. *Le parole sono pietre: Tre giorni in Sicilia*. Turin: Einaudi. English edition: Levi, Carlo. 1958. *Words are Stones: Impressions of Sicily* (trans: Davidson, Angus). New York: Farrar, Straus & Cudahy.
Murray, Carrie A. 2016. *Diversity of Sacrifice: Form and function of Sacrificial Practices in the Ancient World and Beyond* (University of Buffalo. Institute for European and Mediterranean Archaeology. Conference. IEMA proceedings: 5). Albany, NY: State University of New York Press.
Nikolasch, Franz. 1965. Das Lamm als Christussymbol in der frühchristlichen Kunst. Thesis (Habilitationsarbeit), Universität Salzburg.
Pesthy-Simon, Monika. 2017. *Isaac, Iphigeneia, and Ignatius*. Budapest and New York: Central European University Press.
Rasimus, Tuomas. 2012. Revisiting the Ichthys: A Suggestion Concerning the Origins of Christological Fish Symbolism. In *Mystery and Secrecy in the Nag Hammadi Collection and other Ancient Literature: Ideas and Practices: Studies for Einar Thomassen at Sixty* (Nag Hammadi and Manichaean studies: 76), eds. Christian H. Bull, Liv Ingeborg Lied and John D. Turner, 327–348. Leiden and Boston: Brill.
Simeti, Mary Taylor. 1986. *On Persephone's Island: A Sicilian Journal*. New York: Knopf.
Skaggs, Rebecca, and Thomas Doyle. 2009. Lion/Lamb in Revelation. *Currents in Biblical Research* 7 (3): 362–375.
Stroumsa, Guy G. 1992. The Early Christian Fish Symbol Reconsidered. In *Messiah and Christos: Studies in the Jewish Origins of Christianity*, Presented to David Flusser (Texte und Studien zum Antiken Judentum: 32), eds. Ithamar Gruenwald, Shaul Shaked, and Gedaliahu G. Stroumsa, 199–205. Tübingen: J.C.B. Mohr (P. Siebeck).
Tarantino, Maria, and Sabina Terziani. 2010. A Journey into the Imaginary of Sicilian Pastry. *Gastronomica* 10 (3): 45–51.
Terrin, Aldo N. 2019. *Il pasto sacrificale: La violenza nelle religioni* (Pellicano rosso. Nuova serie: 295). Brescia: Morcelliana.
Van Straten, Folkert T. 2016. *Hierà kalá: Images of Animal Sacrifice in Archaic and Classical Greece* (Religions in the Graeco-Roman World: 127). Leiden: Brill.

Massimo Leone is Tenured Full Professor ("Professore Ordinario") of Philosophy of Communication, Cultural Semiotics, and Visual Semiotics at the Department of Philosophy and Educational Sciences, University of Turin, Italy, Vice-Director for research at the same University, and part-time Professor of Semiotics in the Department of Chinese Language and Literature, University of Shanghai, China. He has been visiting professor at several universities in the five continents. He has single-authored twelve books, edited more than thirty collective volumes, and published more than five hundred articles in semiotics, religious studies, and visual studies. He is the chief editor of Lexia, the Semiotic Journal of the Center for Interdisciplinary Research on Communication, University of Turin, Italy (SCOPUS). He is the winner of a 2018 ERC Consolidator Grant, the most prestigious research grant in Europe.

Chapter 4
Phenomenology of a Symbolic Dish: What *Su Porceddu* Teaches Us About Food, Meaning, and Identification

Franciscu Sedda

Abstract How can we analyse a symbolic dish? Which kind of semio-political questions should we consider in such an analysis? Which cultural categories might be useful to study these socio-historical processes and the related forms of life? To answer these questions, we will deal with a specific case study: su porceddu, the roasted suckling pig that represents Sardinia's contemporary symbolic dish. At one level, the analysis allows recognising some relevant issues for Sardinian culture in the broader context of Mediterranean history: first, the dishes' fraught and varied meanings reflect Sardinian traditional, local, regional, or national identities throughout time. At another level, the general categories of continuity/discontinuity, one's own/someone else's, knowledge/flavour, and memory/forgetfulness assist in analysing the meaning of food. This relationalist approach highlights the notion of (un)translatability as a key cultural and alimentary process, also allowing us to look at food consumption and description as embodied forms of and powerful tools for self-consciousness.

Keywords Semiotics · Food studies · National dish · Cultural history · Sardinia

4.1 Introduction

In this essay we will use the history of *su porceddu*, the symbolic dish of contemporary Sardinia, to explore some relevant issues concerning food, meaning and identity.

More specifically, in reconstructing the semio-historical phenomenology of *su porceddu* we will employ five key relations:

(1) The nexus between *cutting* and *deconstructing*: the way food mobilises and transforms common and scientific knowledge.
(2) The nexus between *continuity* and *discontinuity* of the values associated with food: a relation that gets complicated precisely when the presence of a dish in the imagination and the practices of a community sink in very ancient times.

F. Sedda (✉)
Department of Literature, Languages, and Cultural Heritage, University of Cagliari, Cagliari, Italy
e-mail: fsedda@unica.it

S. Stano and A. Bentley (eds.), *Food for Thought*, Numanities - Arts and Humanities in Progress 19, https://doi.org/10.1007/978-3-030-81115-0_4

(3) The nexus between *one's own* and *someone else's*: the way in which the polemical dimension of culture links certain dishes to specific places, to the point of constructing deep correlations between eating and identity.
(4) The nexus between *knowledge* and *taste*: how politically valorised scientific discourse can aggregate naturalistic, economic, political, and gustatory food "data", creating a form of cultural systematisation and self-description.
(5) The nexus between *memory* and *forgetfulness*: how the destruction of cultural memory can provide an opening that allows a symbolic dish to take on diverse meanings.

4.2 Cutting or Deconstructing? On the (Apparent) Opposition Between Common and Scientific Perception of a Symbolic Dish

Antonio Gramsci, the most famous Sardinian in the world, preferred lamb to suckling pig. The fact is witnessed by an impassioned and sagacious defence of the Sardinian lamb against the injustices that this suffered in Turin in 1919, which seems so much an autobiographical tale, but above all by a poignant letter from the prison addressed to his mother in 1931 (Gramsci 1994: 67, 289). Just a matter of individual taste? Probably. Or maybe the roasted suckling pig, the main character of modern Sardinian foodsphere (surely in the eyes of the foreigners, as we will see), reminded him too much of those banquets that infuriated him in his youth: the "illustrious Italian"[1] arrives on the island and eats as much as he can, taking advantage of the hospitality of the Sardinians, after having measured the skulls of the unfortunate returns to the mainland, "pose[s] as Christopher Columbus" and decrees that the Sardinians are mentally ill, an inferior race to be treated as a colony (Gramsci 2008: 51–53). It is the time when, as Gramsci wrote to his wife Giulia Schucht in 1924, he thought that it was necessary to fight for the national independence of Sardinia (Gramsci 1994: 58).

In short, it is difficult to underestimate the existential and political value that a meat-centred banquet can have in Sardinia. It is equally difficult to avoid recognizing the sociality that is generated around food, especially in the collective meaning of a symbolic dish, proving the famous Gramscian idea that all are intellectuals even if only a few perform the function of intellectuals.

Su porceddu serves this purpose. When faced with the dish at a meal, it seems that intellectuals often play a different (language) game others might not. While the common person proceeds to literally cut into the piglet and enjoy it, accepting as a given its origins and meanings, the intellectual poses as his/her task to metaphorically deconstruct the piglet and make its consumers reflect on the historical deceptions and social injustices or environmental issues, which are summoned and even intensified

[1] From here on, when an English translation of the cited works is not provided in the bibliography, the translation has to be considered the Author's.

by the ephemeral and contingent enjoyment of the dish. This creates perhaps the most guilty enjoyment of all, as the meal produces the oblivion of the evils of the world which it itself testifies.[2]

So, cutting or deconstructing? The conflict seems unsolvable. Yet both gestures are related to the commonly understood notions of chopping and breaking up, and share a linguistic and semantic root. They are both *analytical gestures*.

Furthermore, they have the habit of intertwining and intersecting. Is it not perhaps useful for the common person to talk about the dish he/she is eating, its preparation, taste, story, and values? And is it not even the most pedantic intellectual's habit to forget all of the so-called "critical postures" as soon as he/she sits down at the table with the other diners, rediscovering the experience of eating instead of focusing on abstract global concepts of food, identity? On balance, the two are less distinct and distant than they may think.

Yet the conflict is not resolved. The common person sits at the table. The intellectual sits at the desk. The first tastes. The second describes. Moreover, the former perhaps does not want too many problems. The intellectual, on the other hand, is usually looking for them. In this case the latter must address and justify intellectually the notion of "primordialism" (Appadurai 1996), the idea that nations are ancient naturalised phenomena and that people with common culture will eventually coalesce as such. One can argue that *su porceddu* has such an ancient legacy. Given that the consumption and self-representation of *su porceddu* dates back to the Bronze Age, it is impossible intellectualize this dish (and by extension, other symbolic dishes) as "fake" and lacking historical authenticity, or arguing that "traditional" cuisine is only possible after the *Columbian exchange*[3] with the importation of an array of previously unknown and unexpected foods.

A triumph of common sense? Not exactly. The intellectual can reclaim his/her vocation by analysing more to comprehend better: showing how the meaning of the dish changed over time, how it became linked to more general socio-political dynamics, and how became what it is and what it means today. In short, we can intellectualise how and why each bite of *su porceddu* can be a portal, open to everybody, to grapple with both potential and the contradictions of history.

---

[2] Think, in this sense, of the debate about the ethical implications and the ecological impact of the breeding and intensive consumption of cattle, pigs and poultry raised by books such as Foer's (2009) or Harari's (2014). See also the warning of the *UN Intergovernmental Panel on Climate Change* which in its latest report on "Climate change and territory" (August 2019) invited to reduce the consumption of meat as a measure to contrast climate change.

[3] On some general dynamics relating to the circulation of food we make reference to Flandrin and Montanari (1996), and Montanari and Sabban (2006). For a semiotic reading, (see Sedda 2016). On the foods arrived from the Americas and their impact in Sardinia, (see Guigoni 2009).

4.3 Continuity or Discontinuity? Contemporary Assessment of the Bone and Bronze Fragments Evidence

The domestication of animals in Sardinia dates to the Neolithic Era (between the 8th and 7th millennium BCE), but given space restraints our reconstruction will start from the Sardinian Bronze Age, an era that corresponds to the development of the so-called Nuragic civilisation, which has its peak in the recent and final Bronze Age (around 1400–900 BCE). Records of this era found in various *nuraghes* (large stone structures) document early animal husbandry: "the highest number of killings [of pigs] falls in the first two weeks of life or within the first month" (Wilkens 2012: 93).

Further detailed archaeological data confirm a strong correlation between the consumption of piglets and the festive, ritual or even more specifically sacrificial dimension. Describing the so-called "Meetings Hut" of the Palmavera *nuraghe* of Alghero, Perra (2018: 104–107) makes reference to a *nuragic banquet* at which elites took part, sat in a circle around a model of a *nuraghe* (or model of the universe, depending on the interpretations) and alongside a ritual brazier. The ensuing feast, consisting of roasted meat (in particular suckling pigs), then legume soups, bread and wine, celebrated themselves, the community, and the cosmos.[4]

Is there confirmed proof of the role of suckling pig in the self-representation practices of the time? To answer this question, we have to look at the production of bronze statuettes in Nuragic Sardinia, and first of all to a fragment with a hand holding a little animal (see Lilliu 1966). Is it really *su porceddu* that speaks to us from the depths of the Nuragic Age?

First, note that our *unidentified pig object* is located at the intersection of the copious series of bronzes defined as *bidders*—that is male and female offering gifts to deities—and that of those who carry animals on their shoulders, in particular one figure with a ram, defined by Giovanni Lilliu, the father of modern Sardinian archaeology, as "devotee with a ram" (Ibid.: 341).

The fundamental difference between the two groups is that the bidders have the hand, generally the right, raised in the typical greeting (also interpreted as a sign of peace) of the Nuragic Sardinians. This sacred-religious sign coupled with an object (often bowls, but also vases and baskets) or food in the other hand makes one think of a ritual offering. In the male figures—bearers of ram and even deer—both hands are instead engaged in holding the animal, although the type of animals and the clothes suggest a moment of devotion. In one case a male figure holds the legs of the ram with the right and greets with the left, complicating the theory of the presence/absence of the raised hand indicates the sacred and votive meaning of the bronze statuette. In

[4] On the interpretation of this banquet in connection with a hypothetical "crisis of hegemony" within the Nuragic (and Mediterranean) society between the thirteen and twelfth centuries BCE, see Perra (2018: 112). Wilkens (2012: 90), assuming a reproductive cycle of the Sardinian pig similar to that reported by the authors of the Roman era, tends to date the climax of the ritual phases in June or December/January.

this case, Lilliu catalogues it as a bidder (religious in nature) (Ibid.: 341–344). The devotional nature of a statue with a dove is even more clear.

Given these premises, it should be noted that in our fragment the figure Lilliu defines as "swineherd" supports the pig with the left hand—not with two hands and not with the right one. This suggests that the right hand was free and was in the classic gesture of greeting. Hence we can draw two suppositions. First, it is less important to define the representations as a swineherd because these small bronze figurines do not define their identity on the basis of *what they carry* but rather define the value of their gesture for *the way they carry what they carry*.[5] Second, our *porceddu* is a participant in a ritual context, with sacred, perhaps even sacrificial, features, a notion given credence by the fact that the location where the bronzes were discovered is the so-called Nuragic sanctuary of Santa Vittoria di Serri containing a court for feasting and altars for ritual sacrifices (Lilliu 1982: 175–176; Zucca 1988).

Moreover, the ensemble of these bronze figurines seems to define a complex *ritualfoodsphere*, containing many elements still to be identified. The fragment of *su porceddu*, however, is one of the few certain foods (along with the mutton and bread) involved in Nuragic ritual offerings. To this, we can also add water, given that the cult of water is central to the Nuragic culture, as well as wine.

Back to our fragment: is it really a piglet? The only element of uncertainty is given by the plasticity of the mantle, with deep herringbone engravings that suggest a "hairy and bristly" specimen, as Lilliu (1966) writes. The same archaeologist, almost to dispel the (unexpressed) doubt that it is a wild boar, concludes by saying that "with this [type of mantle] is intended to represent a mountain pig, with a rough and wild appearance" (Ibid.: 434).[6]

To move forward in the investigation, let us consider two nuragic representations of ships. The first, was found in Etruria, in Vetulonia, in the so-called *Tomba del duce* [The leader's grave]. The second, from which probably the first originates in terms of craftsmanship, was found in Sardinia, in the town called Meana. These two ships demonstrate the important social and economic value of animals for the ancient Sardinians inside and outside Sardinia.

[5] While using all precautions, it should be noted, with Detienne, that in relation to the Greek cult of Demeter, two ways of sacrificing the pig are distinguished: the first is throwing it in a crack in the ground and therefore leaving its meat to rot, the second is slaughtering it and then cooking it. It is only in this latter case that we are in the real sphere of sacrifice and ritual offering. Now, the difference between the two ways of treating the pig is *in the way it is held*: for the tail in the first case, for the leg in the second (with Demeter holding the torch). The latter is exactly the way our bronze piglet is held. It should be noted that is Detienne himself who traces a connection with Sardinia giving another useful starting point with respect to our fragment: "Similar findings [of statues of Demeter with the piglet] also occurred in Sardinia, in the sanctuary of Santa Margherita di Pula (Cagliari), where pigs are morphologically very close to wild boars (…)" (in Detienne and Vernant 1979: 158). Let us stop here with the connections between Sardo-nuragic and Greek sacrificial culture, to be explored elsewhere, but not before having remembered that Pausanias testifies of the donation by the Sardinians (dated to the sixth century BC) of the statue of Sardus Pater, national divinity of the Sardinians, at the temple of Apollo in Delphi (cf. Mastino 2020).

[6] On the general difficulty of distinguishing domestic and wild pigs, (see Perra 2018: 61).

Fig. 4.1 The Nuragic ship found in Vetulonia (by user: Saliko/WikiCommons, (From https://commons.wikimedia.org/wiki/File:Navicella_nuragica,_dalla_tomba_del_duce_a_poggio_al_bello,_vetulonia,_650-625_ac._ca._04.jpg. Last accessed 22 November 2019). 11-Mar-2017, GNU free under a CC BY-SA-3.0 Unported license)

The Vetulonia ship (Fig. 4.1), in particular, is a splendid artifact that demonstrates, according to archaeologists, the intense commercial contacts between Sardinians and Etruscans around the ninth century BCE. It also demonstrates the prestige of Sardinian bronze work, as the ship was one of the objects buried in the grave of a woman—probably a priestess—and a little girl.

It is important to emphasise that this nuragic ship seems almost a Noah's Ark, populated as it is with animals whose presence certainly have symbolic and sacral values, but which can hardly be considered unrelated to what was the sphere of trade at the time.

So which animals are there on the Sardinian Ark? Surely a *deer*, a totemic animal, whose head is the prow of the ship. Certainly two yoked *oxen*—another animal central to the symbolic-material hierarchy—, which not by chance occupy the centre of the ship and connect its sides. Two similar but different animals, not by chance in the same position but on two opposite sides of the boat, follow in order of size. On the left side is a *pig* eating from a bowl, on the right side is a *boar* being attacked by two *dogs*, reproducing a hunting scene. This parallel allows us to notice important elements that signify the difference between the *domestic* and the *wild*: the design of the mantle, the more or less pronounced size of the ears, the shape of the tail. Each of these elements reverberates on the rest of the composition (one could speak of a *language of tails*) and, together with the spatial arrangement of the animals and the reconstruction of the scenes in which they are involved, can help us clarify not only the identification, but above all the meaning of the scene.

Next to the domestic pig is a *ram* apparently about to mate with a *sheep*. On the left side of the ship, but towards the bow, a crouching *dog* keeps an eye on a *hedgehog*. Leaving aside the unrecognisable bumps-figures towards the stern, only two other scenes remain. One places two animals at the neck of the deer, one above and one below. The latter is the one that obviously attracts the most attention, also due to its unrealistic position. What is this animal attached to the neck of the deer? Perhaps a dog, as confirmed by other objects that reproduce a similar scene more clearly. But what about the little animal, the smallest of all, standing above the neck-bow?

Let us stop for a moment and go to the other side. A pet eats from a container, like the oxen and the pig we analysed above. In front of it, we can recognise the silhouette of a bird, perhaps a *duck*. Behind it, an animal that at first sight looks like a crocodile, but is a lurking *fox*, which we can recognise in other bronzes in a similar pose. Here, then, it is the fox that threatens our unsuspecting domestic animal, which feeds on its back, while another face appears above the fox. Is it a dog that intervenes to block the fox a moment before it launches the attack? But the attack on what? A "bull", according to the identification proposed by Lilliu. Is it possible, however, that a fox can threaten a young bull? Is it not a piglet we are tempted to infer? Of course, the shape of the tail is lowered; but does not the domestic pig on the other side also have its tail down, unlike the curled tail of the boar on the other side?

Meana's ship contains fewer animals, is may be due to its deteriorated state of conservation (see Lilliu 1989). On its stern, just behind the yoked oxen, we can see the *swine couple* again. A wild boar and a pig on inverted sides dominate the scene. This time, however, their features are different: no longer greasy, heavy, and basically static but slender, light and snappy. At a first sight, one could easily admit that we are in front of a *little boar*, with a lot of bristly hair, and a *suckling pig*, complete with a curled tail. Yet another one might reply that this is impossible since they are almost as big as the yoked oxen. To solve the clue we can consider that the two are even larger than the yoked oxen. This means that this ancient representation, foregoing any realistic relationship to size, requires us to focus on the pure figures themselves, on those two animals that sail on the waves of culture. So, did we finally get to identify *su porceddu*? Difficult to confirm, but perhaps even more difficult to exclude it.

Assuming that we are in front of three representations of a piglet, we can move to the more interesting question of the relations of *continuity/discontinuity* with our contemporary symbolic dish. If we run the risk of a time ellipsis, we could say that there are traits of *rituality* and *sacredness* embodied by the consumption of *su porceddu* that make the ancient one sounds similar to the contemporary one. Nevertheless, there is a clear shift from an *authoritative-religious* context to a *daily-prosaic* one. Moreover, it would certainly be incorrect to say that the pig or the piglet played the role of *symbolic animal* of the Nuragic civilisation. The *deer* reproduced on the bow of the votive ships and to adorn the handle of the ceremonial swords seems a better candidate to contend the *bull*, symbol of the pre-nuragic civilisation, thus playing the *role of symbolic or totemic animal*. This casts a heavy shadow on the possibility that the suckling pig could be the symbolic dish of ancient Sardinians. And yet it was already part not only of the Sardinian *foodsphere*, but also of the highest rituals of consumption and self-representation. It is enough to sustain the perception of a *long durée* relation between Sardinians and *su porceddu*.

4.4 One's Own or Someone Else's? Eating as Speaking: The Inedible and the Incomprehensible as Forms of the Untranslatable

If we want to grasp the symbolic value and the political-cultural connection between the identity of the Sardinians and the consumption of pork in the Middle Ages, we have to move out of Sardinia.

Let us then descend into a battle of verses between Folgore from San Gimignano (c. 1270—c. 1332) and Cenne de la Chitarra from Arezzo (thirteenth century—1336). Towards the end of the thirteenth century the poet Folgore sang of the beautiful life and virtues of the Tuscan aristocracy in his *Sonetti de' mesi* [Sonnets of the months][7] most likely commissioned by Nicolò de Nisi, lord of Siena. After heralding every month from January to June, Folgore arrives at the sonnet for July. It is important to note that this is the first and only time Siena is mentioned in the entire composition, the city that lies at the heart of the poem and that at that time can be considered the culinary capital of Italy.[8] Here in July sonnet, Folgore paints a musical portrait of a *soiree* in the marvellous Siena countryside, avoiding the heat, having good time, drinking the best wines, eating roasted partridge and pheasants, boiled capons, and sovereign goats. In short, beautiful people having beautiful life.

How does the poet Cenne respond to this *idyll*? With a *parody* that month by month, point by point, reverses the picture sketched and sung by Folgore.[9] What place and dishes does Cenne choose to overturn the pastoral vision of the Sienese July? Instead of Siena, he locates the horrid banquet in the city of Arestano, Instead of partridges and pheasants for the meal Cenne describes despicable "fat and spicy pork meat".[10] *Aristanis*—in Sardinian; *Oristano* in modern Italian—was at that time the capital of Arborea, a Sardinian Kingdom that until the end of the Middle Ages battles to make Sardinia a unified and sovereign kingdom, first against the Pisans then against the Catalans.[11] In short, the passage testifies the stereotypical (and negative) association between pork and Sardinia from an external point of view. More precisely, we should say "piglet" instead of "pork", since until the sixteenth century texts make

---

[7] https://it.wikisource.org/wiki/Sonetti_dei_mesi.

[8] On the mobile culinary geography of Italy, made of cities that pass the scepter of "capital" one to another within a "common circuit of exchanges", (cf. Capatti and Montanari 2016 and Montantari 2010: 14).

[9] https://it.wikisource.org/wiki/Risposta_per_contrarî_ai_sonetti_de%27_mesi_di_Folgore_da_San_Geminiano.

[10] It should be noted that, as the comparison between texts shows, the nutritional or hygienic status of meat is not in question here. Meat is eaten both in the idyll and in its parody. Furthermore, it will be a Sienese physician, Aldobrandino from Siena in thirteenth century, who positively sanctioned the nutritional value that the Middle Ages imagined for meat (Montanari 2012: 72). Rather, it could be noted that in late Middle Ages with the affirmation of city culture pork assumes the value of rural, peasant and traditional food as opposed to beef or sheep (Ibid.: 77). In short, pork consumption is for unsophisticated if not rough people. This dichotomy is likely to act in making sense of the parodic game we are analysing.

[11] For our point of view and a critical bibliography, (see Sedda 2019a, b).

no distinction among the two, referring to what is clearly a roasted suckling pig as simply "a pig" (see Sedda 2015b, 2020).

The whole question could end here and, although substantial, it would probably not seem significant to most people. Instead, it is just an invitation to widen our gaze and realise that meat and Medieval Sardinia are associated in such a profound a way that even today any dictionary of the Italian language reports meaningful traces about it:

> Sardìgna [from *Sardìgna*, ancient variant of *Sardinia*: from the lat. *Sardĭnia (m)*, from *Sărdus* 'Sardo', perhaps with an allusion to the unhealthy air of Sardinia] In ancient times, a place outside the city where carrion and slaughter waste are piled up / (*dial.*) Slaughterhouse department used for the destruction of infected and spoiled meats. (Zingarelli 1996: *ad vocem*)
>
> As a toponym, *Sardigna*, in Florence, denomination of a place on the banks of the Arno, outside the San Frediano gate, where the carcasses of animals (horses, mules, donkeys) were thrown. (Treccani: *ad vocem*)

The poetry of Cenne takes on added meaning. At the moment of the most radical inversion of the Sienese idyll, the poet moves to a land that is already associated with infected meat, its slaughterhouse and its disposal. The reader of the time probably recognises the game that Cenne proposes; if the consumption of the pig in Arestano becomes a *nightmare*, it is not only because the scene describes the scorching heat or the fat and spicy meat, but because it recalls a deeper sense of *unhealthiness*. Wasn't Sardinia also known and identified since the Romans as a malarious, pestilential place, as recorded in *Zingarelli's Dictionary*?

Yet, that is not all. The unhealthiness here is radical and the metaphorical connection is revealing: Sardinia, like its meat, is inedible, indigestible, not metabolisable. Perhaps even deadly. In short, we are in front of a land that, far beyond the purely geographical datum, is considered an *otherness*, an *exteriority*.[12] Sardinia is outside, but outside of where? If we analyse the entire poem by Cenne, we can notice the oppositional association of Arestano, and with it of Sardinia, to a Siena that is at the heart of Italianness: in his parody, in fact, unlike the composition by Folgore, each month presents different parts of the *peninsula*. But it is an *island*, of course, that has the task of marking the leap from *one's own* to *someone else's* and converting *utopia* into *dystopia*.

To confirm the leap beyond the limit of what can be "assimilated" by the Italian semiosphere of the time there is the *homology* between the structure of Cenne's text and that of a coeval text, much more known and authoritative. We are referring to the passage of the *De vulgari eloquentia* (1303–1304) in which Dante Alighieri, in reviewing languages and peoples more or less similar to Italian and Italianess, states:

> We also eliminate Sardinians (who are not Italians, but seem to be associated with the Italians), because they alone appear to lack their own vernacular language and imitate the grammar as the monkeys imitate men: they say *domus nova* and *dominus meus*.[13]

---

[12] On food and otherness, (see Stano 2015).

[13] Fort the original Latin and the Italian translation (see Dante Alighieri 1983: 428–429).

Though we are in front to two different authors and texts with different tones, and with differing issues at stake, the parallel appears evident: whereas in Dante there is a *way of speaking* that is *beyond the limit of the recognisable* and the minimally ordered, which allows to mark its speakers with a trait of *animality*, so in Cenne there is such a *fat and peppered pork meat*—and a banquet so horrid overall—that it is *beyond the limit of the edible* and also of the *thinkable*, such that its dishes end up in the semantic field of the *repugnant*. On one hand, therefore, we have men who, although speakers, become animals; on the other hand, we have animals which, although cooked, become humanly inedible and dangerous. This *otherness*, perceived and described in such a radical way is exactly what enters in the cultural memory through the attribution of the name *Sardigna* to the *place of disposal of infected and spoiled meats*. What can there be in fact more animal-like and repulsive *at the same time*?

In short, *talking and eating are one thing and perform the same story*. The pork meat, like the Sardinian language, becomes the sign of the Sardinians and Sardinia otherness.

4.5 Knowledge or Taste? Or How the Suckling Pig Makes the Nation

> The pig does not awaken any cry of the economy, nor does it impose the nuisance of reforms; indeed, it receives the greatest applause: excellent for feasting, it still offers a singular form, and appearance, so that an Italian finds the Sardinian pig new, and the Sardinian the same with the Italian pig. (Cetti 2000: 108)

Thus begins the part of the discussion that the Jesuit Francesco Cetti, Italian zoologist and mathematician, distinguished professor at the University of Sassari starting from 1765, dedicates to the pig.

The Sardinian pig in Cetti's *Storia naturale della Sardegna*[Sardinia's Natural History], his major work published between 1774 and 1778, which combines zoology, economy, history and politics, occupies the least central place. First of all, it is designated so for its qualities. In the midst of so many imperfect or economically unprofitable Sardinian animals, says Cetti, "Only the pig strictly deserves praise" (Ibid.: 87). And that is why it is *abundant* and *tasty* (Ibid.: 109). Better, Sardinian pig is *tastier*: "The meat is generally firm, and superior in flavor to that of Italy" (Ibid.: 110).

It is difficult at this point to maintain that the pig does not rise to a symbol of distinction, an object of value that embodies—and through the ritual of consumption allows to incorporate—a feeling of self-esteem and collective identification.

> To the extent of goodness and abundance the appetite and the satisfaction among the nation is great [...] and if the vintage goes wrong for any accident, the populace whispers, nor does it claim to see its [skewer] empty. (Ibid.)

Should the Sardinian pig be considered a sensible vehicle of the *imagination* of the Sardinian nation, to say it with Anderson (1983)? Or, echoing Gramsci (1992), as a fundamental part of *national-popular life*, to the point that its lack puts tension in the populace? The hypothesis is not to be discarded too resolutely reducing it to a mere provocation. The roasted pig is at the centre of a more complex ritual of encounter and differentiation. The moment of the feast, with the consumption of meat, dances, poetry in Sardinian language, and the ritual of hospitality of foreigners are intertwined. Their correlation nurtures a process of cultural self-recognition and contribute to that awakening of a *Sardinian national conscience* that will lead to what will be called *Sarda Rivoluzione* [Sardinian Revolution], whose apex is conventionally placed between 1793 and 1796.

It is within this dynamic that the act of *roasting* can be recognised as the *collective passion* of the Sardinians and pork in general, and *porceddu* in particular, can become a *symbol of national diversity*. Indeed, for Cetti the national diversity of Sardinia is so evident that the problem becomes its *origin*, which ends up calling into question a much more radical temporal and spatial alterity, which goes far beyond the borders of the island:

> The roast is the object of the greatest passion among Sardinians, it forms the point of support for every banquet, neither more nor less, than it did in Ithaca or Mycenae. A thousand times, seeing the hospitality and customs of the Sardinian countryside, Homer came to my attention, and it seemed to me to be Telemachus, who was traveling in Greece. The guest without ever having seen you, or known, welcomes you courteously, has horses dismantled, introduces you. The handmaids immediately leave their looms in the vestibules, bustling about making fresh bread; meanwhile the sheep, the suckling pig, comes from the shepherd; it is cut into pieces, and on a spit it is placed on the fire (...). (Cetti 2000: 164)

From Cetti's *happy orientalism*,[14] a point-by-point parallelism ensues between practices of ancient Greece and modern Sardinia that cannot but culminate with the *ballu tundu* [Sardinian circular dance][15] and give foundation to the idea that the initiator of all this was the Greek Iolaus, who "landed with triumph, placed himself in the heart of the island, was the age of civil life, of agriculture, of cities, of a nation" (Ibid.: 165). And ended up being consecrated "Father of the Sardinians" (Ibid.). This would be demonstrated, according to Cetti, by the fact that the Sardinian system of animals is comparable to the Greek one, since the animals came with Iolaus.

Leaving aside the fact that this reconstruction is nowadays considered a myth, what is interesting is this connection between animals and anthropology, between knowledge about the former and the latter. Even more intriguing is that there is a deeper way, rooted in sensuousness,[16] through which the pig reveals its socio-political value.

---

[14] Between eighteenth and nineteenth century Sardinia will be the object of a clearly "orientalising" gaze. Differently from most part of the others foreigners writing about Sardinians' cultural distance from "Europeanness", Cetti does not see the "Easternness" of Sardinians as a sign of barbarousness to be redeemed but as a positive trait of distinction and civilisation.

[15] On the meaning and role of Sardinian dance, (see Sedda 2003) [2019].

[16] For an example of disentanglement of cultural values from sensuousness, (see Harkness 2013).

We have seen in fact how the difference between Sardinia and Italy in Cetti is based on a *diversity of flavour*. Or, even more clearly, on a different and better quality of Sardinian pork compared to the Italian one. The depth of this *taste-valuenexus* is witnessed by an apparently minor fact: the praise of what Cetti calls "marine pig" or "pig fish", *tuna*, in the same *Natural History of Sardinia*.

The point that creates the *equivalence* is not simply that nothing from tuna or pig is ever thrown away.[17] The point is that, according to Cetti, Sardinian tuna *tasteslike pork*. Is it possible? Our palates would have difficulty recognising it and our tongues to admit it. But even if we were witnessing a powerful cultural game—as confirmed by the fact that Cetti boasts "bottarga", dried tuna eggs, calling it "salami"—how can we exclude the possibility that Cetti actually perceived a similarity of taste between pork and tuna? The variability of tastes confirms this. Intellectual and social appreciation can act on the palate, on what we feel through the senses: indeed, generally it does, because as Massimo Montanari has vividly said, "the organ of taste is not the tongue but the brain" (Montanari 2004: 73).[18] In short, if the best flavour of a culture is that of the pig, then the best fish in that same culture will be a "sea pig" and will taste like pork.

This *elevation of an animal/flavour to the meta-level*, its transformation into *a dominant* that spreads its structural (and sensory) traits on cultural material (Jakobson 1971; Lotman 1985: 132–133), demonstrates a *reflective performativity* that allows collectives to define themselves, to create a system, to elevate a portion of themselves to symbolise the whole, in order to achieve a positive sense of unity (see Sedda 2015a).

In short, it is difficult to underestimate all these naturalistic, economic, historical, gustatory evaluations—i.e. this complex web of connections between tastes and knowledge—from the perspective of the cultural and political history of Sardinia while not forgetting the *authoritativeness of the source* and the *scientific natureof the naturalistic discourse*, that is to say, the *hierarchical position* that this type of utterances occupied in the *episteme*, in the system of truth, of the time (see Foucault 1968; Greimas 1970). It is no coincidence that some of the major politicians and intellectuals of the *Sardinian Revolution*, like Giovanni Maria Angioy and Matteo Luigi Simon, reference Cetti's work to explain the origins of the Sardinian nation and give further foundation to the feeling of national diversity of the Sardinians from the Italians at the end of the eighteenth century.

---

[17] And like pork, tuna bleeds. Indeed, it is the only fish to do it. So much so that for the Greeks, for example, it is the only fish that can be sacrificed to the gods (cf. Durand in Detienne and Vernant 1979: 148–149).

[18] For a semiotic approach to food, which addresses similar issues, (cf. Marrone 2012 and Marrone 2015).

4.6 Memory or Forgetfulness? Semiopolitics of the Contemporary Piglet

One thing about which the intellectual is (perhaps) right is that much of what is recovered through the analysis of cultural history—knowledge unknown or misperceived by those who taste the piglet—often projects on the past more than what it has value for them in the present.

If this happens, however, it is because of processes of transformation/destruction of memory (Lotman and Uspensky 1973). In fact, what we have not said is that each of the phases analysed has been followed by moments of radical denial of the experiences and meanings associated with them. The *effect of continuity* that the succession of parts of our essay produces is in many ways distorting. In fact, it does not correspond to a linear historical dynamic, in which each phase tackled draws explicit or direct nourishment from the previous one.[19] Moreover, the events following the repression of the *Sardinian Revolution* opened the way to further fractures of the historical memory of the Sardinians that we cannot investigate here: they are so deep that semiotics, intended as a *therapeutic of the social*, as Greimas (1987) once defined it, can contribute to heal them only to the extent that it is connected to wider transformations of society.

It is the presence of these *traumas*, this cancellation of connections and values that, to give an example, can allow a political action marked by the most vigorous Italian nationalism, like that of recent populist Italian minister Matteo Salvini. At his arrival in Sardinia, launching his party campaign for the Sardinian autonomic elections, in a usually xenophobic, pro-Italians and anti-immigrant speech Salvini mobilised a symbol that has appeared throughout the centuries more or less explicitly, as a sign identifying Sardinians as distinct from Italians: "For those who now govern Sardinia, the island must be repopulated with [Muslim] immigrants, who then cannot eat the suckling pig, and will prevent us from doing the crib and hang the crucifix: stay at your home!".[20]

The traumas of history, resulting in the twentieth century emergence among Sardinians of the paradigm of Sardinia as a *failed nation* (analysed elsewhere),[21] prevents Salvini from being heir to those "illustrious Italians" stigmatised by Gramsci.

The historical distinction accrued from the suckling pig, detached from any Mediterranean matrix, from any Sardinian national and socially progressive memory, reduced to pure folkloric object, can be easily mobilised to activate a *conservative* and *regionalistic* form of identity pride that is compatible, if not functional, with the assimilation of Sardinian practices into the framework of an Italian nationalist

[19] We have outlined, starting from the sixteenth century, a more complete history in Sedda (2015b); here we have shown in more detail how starting from the nineteenth century the pig turned into a "local dish". A global history of *su porceddu* can now be found in Sedda 2020.

[20] http://www.lanuovasardegna.it/regione/2018/11/22/news/matteo-salvini-a-olbia-da-marzo-la-lega-governera-la-sardegna-1.17488597.

[21] We refer to (Sedda 2002, 2003, 2017), [2019].

discourse that aims to distinguish an Italian-Christian *We* from an immigrant-Muslim *Them*. In this context, the Salvinian phrase can even be read by many Sardinians as a form of valorisation: both because it takes the suckling pig "at the national level", that is at the meta-level of the contents of the Italian political discourse, perceived by the contemporary majority of the inhabitants of the island as the place of production and sanction of *true* and *higher* values, and because it attributes to the symbolic dish a sort of sacredness placing it in *equivalence* with traditional-religious facts such as the crib and even the crucifix.

This last example shows that the roast suckling pig, precisely because it is mobilised and mobilisable within contemporary conflicts, is in effect one of the central symbols of contemporary Sardinia. At the same time, it shows how the values historically connected or connectable to it are largely forgotten. This semio-cultural analysis can contribute to their remembrance. At the same time, it makes them available both for a critique of contemporary political discourses and for alternative translations in the future: all the while allowing greater awareness of the complexity of the historical-political development and of the value that food, the relations it incorporates, can assume inside the becoming of history.

References

Anderson, Benedict. 1983. *Imagined Communities*. London: Verso.

Appadurai, Arjun. 1996. *Modernity at Large. Cultural Dimensions of Globalization*. Minneapolis-London: University of Minnesota Press.

Capatti, Alberto, and Massimo Montanari. 2016. Italia. In *Storia e geografia dell'alimentazione, Vol. 2: Cucine, pasti, convivialità*, eds. Massimo Montanari and Françoise Sabban, 804–825. Torino: UTET.

Cetti, Francesco. 2000. *Storia naturale di Sardegna*, eds. Antonello Mattone and Piero Sanna. Nuoro: Ilisso. Original edition: Cetti, Francesco. 1774-1778. *Storia naturale di Sardegna*. Sassari: Piattoli.

Dante Alighieri. 1983. *Opere minori. Volume Primo. Vita nuova, De vulgari eloquentia, Rime, Ecloge*, eds. Giorgio Barberi Squarotti, Sergio Cecchin, Angelo Jacomuzzi, and Maria Gabriella Stassi. Torino: UTET.

Detienne, Marcel, and Jean-Pierre. Vernant. 1979. *La cuisine du sacrifice en pays grec*. Paris: Gallimard.

Flandrin, Jean-Louis., and Massimo Montanari, eds. 1996. *Histoire de l'alimentation*. Paris: Fayard.

Foer, Jonathan Safran. 2009. *Eating Animals*. New York: Little, Brown and Company.

Foucault, Michel. 1968. *L'archeologie du savoir*. Paris: Gallimard.

Gramsci, Antonio. 1992. *Folclore e senso comune*. Roma: Editori Riuniti.

Gramsci, Antonio. 1994. *Vita attraverso le lettere*, ed. Giuseppe Fiori. Torino: Einaudi.

Gramsci, Antonio. 2008. *Scritti sulla Sardegna*, ed. Guido Melis. Nuoro: Illisso.

Greimas, Algirdas Julien. 1970. *Du Sens*. Paris: Éditions du Seuil.

Greimas, Algirdas Julien. 1987. Algirdas Julien Greimas mis à la question. In *Sémiotique en jeu. À partir et autour de l'œuvre de A. J. Greimas*, eds. Michel Arrivé and Jean-Claude Coquet, 301–330. Hàdes-Benjamins: Paris-Amsterdam-Philadelphia.

Guigoni, Alessandra. 2009. *Alla scoperta dell'America in Sardegna. Vegetali americani nell'alimentazione sarda*. Cagliari: AM&D.

Harari, Yuval Noah. 2014. *Sapiens: A Brief History of Humankind*. New York: Harper Collins.

Harkness, Nicholass. 2013. Softer soju in South Korea. *Anthropological Theory* 13 (1/2): 12–30.

Jakobson, Roman. 1971. The Dominant. In *Readings in Russian Poetics: Formalist and Structuralist Views*, eds. Ladislav Matejka and Krystyna Pomorska, 82–87. Cambridge (Mass.): Harvard University Press.

Lilliu, Giovanni. 1966. *Sculture della Sardegna nuragica*. Cagliari: La Zattera. New edition: Lilliu, Giovanni. 2008. *Sculture della Sardegna nuragica*. Nuoro: Illisso.

Lilliu, Giovanni. 1982. *La civiltà nuragica*. Sassari: Delfino.

Lilliu, Giovanni. 1989. Meana dalle origini all'alto medioevo. In *Meana. Radici e tradizioni*, ed. Giovanni Lilliu, Anacleto Luciano, Giuseppe Luigi Nonnis, Maria Antonia Sanna, Giancarlo Sorgia, 27–100. Cagliari: Stef.

Lotman, Jurij M., and Boris A. Uspenskij. 1973, *Tipologia della cultura*. Milano: Bompiani.

Lotman, Jurij M. 1985. *La semiosfera*. Venezia: Marsilio.

Marrone, Gianfranco. 2012. *La cucina del senso. Gusto, significazione, testualità*. Milano: Mimesis.

Marrone, Gianfranco (ed.). 2015. *Buono da pensare. Cultura e comunicazione del gusto*. Roma: Carocci.

Mastino, Attilio. 2020. L'iscrizione latina del restauro del tempio del Sardus Pater ad Antas e la problematica istituzionale. In *Il tempio del Sardus Pater ad Antas*, ed. Raimondo Zucca, 199–239, Roma: Bretschneider.

Montanari, Massimo. 2004. *Il cibo come cultura*. Roma-Bari: Laterza.

Montanari, Massimo. 2010. *L'identità italiana in cucina*. Roma-Bari: Laterza.

Montanari, Massimo. 2012. *Gusti del Medioevo. I prodotti, la cucina, la tavola*. Roma-Bari: Laterza.

Montanari, Massimo, and Françoise Sabban (eds.). 2006. *Storia e geografia dell'alimentazione, Vol. 1: Risorse, scambi, consumi*. Torino: UTET

Perra, Mauro. 2018. *Alla mensa dei Nuragici. Mangiare e bere al tempo dei nuraghi*. Sassari: Delfino.

Sedda, Franciscu. 2002. *Tracce di Memoria*. Cagliari: Fondazione Sardinia.

Sedda, Franciscu. 2003. *Tradurre la tradizione. Sardegna:* su ballu, *i corpi, la cultura*. Roma: Meltemi. New edition: Sedda, Franciscu. 2019. *Tradurre la tradizione. Sardegna:* su ballu, *i corpi, la cultura*. Milano: Mimesis.

Sedda, Franciscu. 2015a. Semiotics of Culture (s). Basic Questions and Concepts. In *International Handbook of Semiotics*, ed. Peter Pericles Trifonas, 675–696. Berlin: Springer.

Sedda, Franciscu. 2015b. *Su porceddu*. Breve storia culturale del piatto simbolo della Sardegna fra XVI e XX secolo. *Lexia* 19-20: 133–152.

Sedda, Franciscu. 2016. Glocal and Food: On Alimentary Translation. *Semiotica. Journal of the International Association for Semiotic Studies* 211: 105–126.

Sedda, Franciscu. 2017. L'aspetto della nazione. Divenire, tempo e storia in un caso di negazione nazionale. *Lexia* 27–28: 527–553.

Sedda, Franciscu. 2019a. *Insula, natio* o *republicha*? La Sardegna in due testi fondatori di epoca medievale. In *Isole settentrionali, isole mediterranee. Letteratura e società*, eds. Mariella Ruggerini, Veronka Szöke, and Morena Deriu, 345–374. Milano: Prometheus.

Sedda, Franciscu. 2019b. Alle radici de *sa Battalla*: l'emersione della sardica natio come concetto e soggettività (1353–1359). In *Sanluri 1409. La battaglia per la libertà della Sardegna*, ed. Franciscu Sedda, 41–88. Cagliari: Arkadia.

Sedda, Franciscu. 2020. *Su porceddu. Storia di un piatto, racconto di un popolo*. Cagliari: L'Unione Sarda.

Stano, Simona. 2015. *Eating the Other: Translations of the Culinary Code*. Newcastle upon Tyne: Cambridge Scholars Publishing.

Treccani, Vocabolario online della lingua italiana, http://www.treccani.it/vocabolario/.

Wilkens, Barbara. 2012. *Archeozoologia. Il Mediterraneo, la storia, la Sardegna*. Sassari: Edes.

Zingarelli, Dizionario della lingua italiana, ed. 1996.

Zucca, Raimondo. 1988. *Il santuario nuragico di Santa Vittoria di Serri*. Sassari: Delfino.

Franciscu Sedda is Associate Professor of Semiotics at the University of Cagliari where he teaches General Semiotics and Cultural Semiotics. He served as vice president in the Italian Association of Semiotic Studies and is currently the Secretary of the Italian Society of Philosophy of Language. Professor Sedda was awarded the "Sandra Cavicchioli" prize, chaired by Umberto Eco, for the best MA thesis in semiotics in 2000–2001. The work is published under the title *Tradurre la tradizione* [*Translating Tradition*] (Rome 2003, new edition Milan 2019). Other volumes include: *La vera storia della bandiera dei sardi* [*The True History of Sardinians' Flag*] (Cagliari 2007), *Imperfette traduzioni. Semiopolitica delle culture* [*Imperfect translations. Semiopolitics of cultures*] (Rome 2012), *Roma. Piccola storia simbolica* [*Rome. A Short Symbolic History*] (with Paolo Sorrentino, Rome 2019). Edited volumes include: *Glocal. Sul presente a venire* [*Glocal. On present to come*] (Rome 2004), Jurij M. Lotman, *Tesi per una semiotica delle culture* [*Thesis for a Semiotics of Cultures*] (Rome 2006), *Isole. Un arcipelago semiotico* [*Islands. A Semiotic Archipelago*] (Milan 2019).

Chapter 5
Food Heritage, Memory and Cultural Identity in Saudi Arabia: The Case of Jeddah

Cristina Greco

Abstract The Kingdom of Saudi Arabia (KSA) is quickly becoming a cultural hub, and experiencing changes in patterns of culture and social behaviour. As a result, KSA is seeking to develop and expand its non-oil economy, transitioning into a destination more attractive for visitors and investors. In line with its "Saudi Vision 2030", a strategic framework to reduce Saudi Arabia's dependence on oil, diversify its economy, and develop public service sectors such as health, education, infrastructure, recreation, and tourism (https://vision2030.gov.sa/), the country is developing different sectors, revalorising and revitalising its culture and foreseeing a form of modernisation based on a spirit of innovation. This chapter aims to explore how Saudi cultural identity is expressed and negotiated in different discourses regarding food heritage, and how *hybridity* generates an identity crisis between diversity and integration. In the light of upcoming transformations, how is Saudi cultural identity expressed, constructed, and negotiated in different discourses? Societal changes can occur either as progressive and linear transformations or as abrupt bursts of activity that could affect both the social environment and the country's image. Change and transformation can be newly built from scratch, or can be a revitalisation of existing identity. Through an exploration of textual materials and methods such as observation and interviews, the research explores the meaning of food tradition and its representation through individual and collective memory.

Keywords Food heritage · Cultural identity · Cultural memory · Hybridity · Self-determination

5.1 Introduction

Many studies have been conducted on traditional food and cultural identity (Stano 2015; Md Ramli et al. 2016; Becuţ and Lurbe i Puerto 2017), but little academic work has been done on food and identity in the Kingdom of Saudi Arabia (KSA). Given

C. Greco (✉)
Advertising Communication, Kingdom of Saudi Arabia, University of Business and Technology, Jeddah, Saudi Arabia
e-mail: c.greco@ubt.edu.sa

S. Stano and A. Bentley (eds.), *Food for Thought*, Numanities - Arts and Humanities in Progress 19, https://doi.org/10.1007/978-3-030-81115-0_5

the rapid transformations currently taking place, however, it has become imperative to conduct such a study. Social changes, for example, include the issuing of tourist visas; the removal of bans, including the ban on women driving and women travelling without guardianship; the old emigration phenomena; and the role of mass media. All have an impact on food consumption, production, distribution, preparation, perception and practices. The recent interest in and development of a local food heritage reveals the active role of history through which to understand Saudi culture.

As Barthes states in his introduction to Brillat-Savarin's *Physiologie du Goût* (1838), food "is, in short, a total social fact around which a variety of metalanguages can be gathered: physiology, chemistry, geography, history, economics, sociology, and politics (today, we could add symbolism)" (Barthes 1975: 32).

This chapter aims to explore Saudi food identity and notions of authenticity, as expressed through the discrepancies between native inhabitants' and expatriates' (temporary inhabitants' or tourists') perceptions. In this analysis, food embodies the contradictions, peculiarities and local aspects of Saudi Arabia, and reveals the similarities as well as diversity of cultures. In the light of these variations, this study considers the role food plays in representing and preserving the cultural identity. Which aspects of food, and its related habits and rituals configurations, make it an effective tool for memory? What kind of memories are inscribed, consciously or unconsciously, in food consumption, and what is implicated in a forgetting process?

5.1.1 Food, Memory and Identity

In considering the relationship between food, memory and identity, this chapter attempts to outline a conceptual framework that describes how food across cultures is inscribed in memory and identity, as well as the aspects and practices implicated in conscious and unconscious processes of remembering and forgetting. The following pages examine how we construct a cultural food identity based on what we recollect as "traditional", and "local," often forgetting that it belongs to other *semiospheres* (cf. Lotman 1984). In this sense, the chapter explores the dialectics between remembering and forgetting. These occur through a different range of processes, collective and individual, which constitute different faculties in marking, inscribing and interpreting the past. These processes go under the concept of memory, which could be inclusive of a literal meaning or of a more metaphorical use of the term (Avieli 2013).

When related to food identity, the concept of memory has been treated in several studies on food in semiotics and related fields, including sociology and anthropology (Appadurai 1988; Matta 2013). We will consider the concept of identity as developed by Lévi-Strauss (1977), that is to say, of "an identity built on a relational outlook, in which the other is integral part of the identity" (Boero and Greco 2018: 23). Furthermore, the meaning of identity establishes the presence of the other as an external element (Ricoeur 2006), even though the recognition of the other is missing and the cannibalization takes place transforming the foreigner in a member without leaving clear traces of such a translation process.

5.2 The Research Perspective

This chapter is part of an ongoing research project, made possible because of the new opportunities for development as expressed in Saudi Vision 2030. It adopts a multidisciplinary approach within the semiotics of culture, and employs the qualitative methods of focus group, observation and in-depth interviews.

The research considers two perspectives. The first one is the representation of food identity by the public sector and destination marketers through media and information sources, and by the consumers through practices. The second one concerns food authenticity as perceived by autochthones (original inhabitants) and expatriates (non-citizen inhabitants and tourists), which reflects their perception and awareness of traditional food. The aim is to integrate food identity and traditional cuisine into the overall set of signifying practices. To explore this issue, conceptual frameworks have been developed on the determinants of food identity influenced by abrupt changes as well as gradual development (Lotman 1992), cultural memory and authenticity. In particular, *authenticity* is here considered as a quality socially ascribed and produced by the discourse about food (Handler and Linnekin 1984). One goal of this study is to introduce a first reflection on the meaning of authenticity in contemporary Saudi Arabia.

The research began with an exploratory phase in Jeddah in April 2019, consisting of participant observations, as well as informal and semi-structured interviews with different types of actors,[1] involving a group of students at Jeddah College of Advertising, University of Business and Technology. Focusing on the framework of traditional food and destination, informal interviews were conducted in Al-Balad.[2] Autochthones, consumers and vendors provided the first ethnographic data, while the exploration of the media discourse, the observation of rituals related to food, and in-depth interviews provided a remarkable view of what is considered "traditional", what is remembered, and what forgotten. In other words, how people ate and continue to eat, live, and perceive how "local food" is different from "traditional food" and how these define and redefine their cultural identity and the meaning of authenticity. The research design began by considering the historical and geographical aspects of the city as well as the socio-cultural *milieus*, tracing the representation of the food identity through public and private channels, website and newspaper articles, and the habits and rituals through observation, interviews and projective techniques. In particular, the corpus consisted of the institutional website of tourism in KSA (www.visitsaudi.com), two national newspapers (*Saudi Gazette* and *Arab News*), the historical site of Al-Balad, the semi-structured interviews and the projective techniques insights used in the in-depth interviews to uncover deeper perception. Ten respondents of autochthones and expatriates were interviewed, while an early focus

[1] The semi-structured interviews, preceded by observation and informal interviewing, have been administered in Al-Balad to key informants with personal experience, perceptions, beliefs and a specific knowledge of the population and topic of interest.

[2] Al-Balad is the old city of Jeddah.

group[3] consisted of seven undergraduate students, who were originally from different regions of Saudi Arabia. For the projective techniques, respondents were asked for what name and adjective come to mind upon certain pictures of Saudi dishes.

5.3 Saudi Arabia's Cultural Identity Between Gradual Development and "Cultural Explosions"

Saudi Vision 2030's aim is to broaden and diversify the country's economic sector and invest in new projects over the next decade. This development plan is to be accompanied by social and economic reforms to reduce the level of dependence of the country on the oil industry.[4]

The development plan includes the expansion of crucial industries, such as entertainment, tourism, culture and higher education. However, in addition to the economic transformations, there seems to be the determination to provide a narrative of KSA's culture for the outside world, featuring developed explanations of *cultural identity* and *self-determination*. In line with this cultural and socio-economic plan, Saudi Entertainment Ventures is promoting national cultural events, seeking to increase the involvement of Saudi families and encouraging their participation and spending in the entertainment sector. Examples of these events are the concerts at King Abdullah Economic City (Riyadh and Jeddah 2018, 2019) and at Tantora festival (2018–2019) in historic Al Ula; the Italian Super Cup (Jeddah 2019); the project of Neom and the role of artificial intelligence in building a new city; and the Red Sea Project, a nature preserve intended to offer coral-reef diving on several dozen islands.

These changes are not occurring as a gradual development, but as abrupt cultural bursts of activity, affecting the social environment and the country's image. One of these is the opening of cinemas in Riyadh and Jeddah, a removal of a thirty-five-year ban, with VOX leading the market and MUVI as the local company. Another change is the end of the driving ban for women, which allows them to drive for the first time in the country's history, and the end of the ban on women travelling without guardianship. The current situation of rapid technological, cultural and socio-economic change goes hand in hand with an increasing globalisation, affecting different types of consolidated systems and influencing the development of educational and training systems. This requires a new strategic plan for the construction of a well-recognised identity for the cities that seek to become attractive tourist destinations. All of this involves ideas and processes of both building something new from scratch, and also of renewing an existing culture to preserve and highlight its uniqueness and reactivate the cultural memory of the places. Therefore, it is necessary to

[3] The focus group was conducted in the first phase of the exploratory research, during the field observation period (April–May 2019), while the interviews were conducted between September 2019 and April 2020.

[4] https://vision2030.gov.sa.

start considering not only the historical places and architecture, the ancient monuments and the old storytelling, but also food culture and rituals, and more generally everything the city needs to document and preserve from its cultural heritage.

With the new expectations of Saudi Arabia's travel and tourism sectors, which opened on 27 September 2019, and the number of international visitors expected to rise progressively until 2023 (presuming the eventual cessation of the coronavirus pandemic), Jeddah started a design process of the city that includes new development of its sea side and north side, basic infrastructure enhancements, renewal of the old city, and a variety of entertainment structures and projects.

In this respect, we first explore the historical and geographical context, focusing on Jeddah as a case study, to highlight its characteristics. We then examine the definition of "local food" as recognised by people and how through food memory is collected, stored, preserved and communicated, in a country experiencing an uncertain transition of its cultural and economic institutions. We discuss two types of processes, defining the variance between them as the distinction between abrupt change and gradual development, to reference Lotman in *Culture and Explosion* (1992).

5.3.1 Saudi Arabia. Bedouins' Diet and Food

The history of the Kingdom of Saudi Arabia's society and politics, from the Ottoman Empire suzerainty, the oasis emirs and tribal confederations, the role of the caravan routes, to the emerging State and the Gulf War (Al-Rasheed 2002), is a complex scenario that needs a specific focus.

This is neither the place, nor the research's aim, to discuss the historical events of Saudi Arabia. We will focus instead on some of the historical aspects and contemporary facets of Saudi Arabia and, more specifically, Jeddah, to address the tension between food, memory and the perception of its authenticity in this specific case, where food provides a space in which to explore the complexities of the cultural identity construction.

Given these limitations, with respect to food consumption, the origin of Saudi Arabia can be traced to the history of Bedouin tribes of nomadic sheep and goat herders. This group maintains an influence on the Saudis' current diet and plays a role in the notion of food authenticity. From a nutrition standpoint, the average meal of nomads who still inhabit Saudi Arabia is less complex compared to that of the urban Saudi Arabian people who represent the plurality of the population nowadays. Yet, the essential food components are the same, as fava beans, wheat, rice, yogurt, dates and chicken are staples for all Saudi Arabian people.[5] Conventional food items such as dates, *Fatir* (flatbread), *Arikah* (bread from the south-western part of the country) and *Hawayij* (a spice blend) are still eaten by Saudis, in spite of the fact that most of the population have settled in cities and towns. Given the pervasive practice of Islam (and its state-sanctioned status), Saudi food culture is intensely

[5] Saudi Arabia has over 18 million date palms that create 600 million pounds of dates each year.

characterised by dietary limitations on alcohol and pork consumption, as the dishes must conform to what is deemed permissible or *ḥalāl*. Islamic law plays an important role in organising cooking practices and food preparation. For example, the animals must be butchered in a specific way and blessed before consumption. According to the Food and Agriculture Organization of United Nations (FAO), "Halal Food means food permitted under the Islamic Law", and should fulfil a number of conditions, such as that of not containing characteristics considered to be unlawful according to Islamic Law. It refers to diverse categories that must be considered in the classification of lawful food, such as Food and Animal Origin, Food and Plant Origin, Slaughtering, Preparation, Processing, Packaging, Transportation, Storage, and Labelling (Codex Alimentarius Commission, 22nd Session, 1997).[6]

In terms of the country's development, the oil concession in 1933 (Al-Rasheed 2002), and the related increase in the income allowed Saudi Arabia to reorganise the society and improve other sectors such as agriculture. The result for the country was a well-developed national production of dairy products and vegetables. The per capita annual consumption of chicken is particularly large; in fact, Saudis rank as the largest consumers of broiler chickens in the world, while lamb is traditionally served over the holiday feasts and to honour guests.[7] Camel milk, as well as sheep or goat milk, has long been the main drink of the Bedouin diet, and dairy products are still top choices for the population. Yogurt is considered an essential component: it can be eaten alone or accompany dishes in forms of sauces and drinks, such as *lassi*. Other essential elements of the nomadic diet, equally eaten by Saudis, are *Fatir*, mentioned above, and *Kimaje*, types of bread used as a substitute for fork and spoon.[8]

5.4 Jeddah "Ghair"

Jeddah, located in the Hijazi region, is the second largest city in KSA and the largest in Makkah Province with an estimated population of over four million.[9] Since it was established almost 2000 years ago as a fishing hamlet, Jeddah has played a crucial geographical and political crucial role in the Arabian Peninsula. Located along the Red Sea, the city is not only a commercial centre, but also a hub for maritime activity. The proximity to Mecca and Medina, Islam's holy places, makes it an essential midpoint for Muslim pilgrims flocking to perform the rites of Hajj and Umrah. Over time, Jeddah, known as "the bride of the Red Sea", has become a cosmopolitan hub and a crossroads for people from all over world, exchanging experiences and knowledge, and thus enriching the local community.

---

[6] The term *ḥalāl* includes several conditions and criteria to recognise the sources not considered lawful according to the Islamic law (www.fao.org, CAC/GL 24–1997[27]).

[7] Saudis eat an average of 88.2 lb of chicken per person per year. With regard to the import of local products, Saudi Arabia stood as world's largest importer of live sheep (www.foodbycountry.com).

[8] www.foodbycountry.com; www.tasteatlas.com.

[9] Jeddah is considered the 100th largest city in the world by land area.

In early times, various groups, including the Tulunids and Abbasids, ruled the city. A wall was raised in the 1200s, part of which still exists in Jeddah, to secure the city against the Portuguese's assaults. The wall fell in the 1500s during the Ottoman Empire, but was later rebuilt to protect the city.[10] During World War I Jeddah rebelled against the Empire, and later was conquered by Ibn Saud, becoming part of the Al-Saud dynasty. While a fire in the 1980s damages the city walls and old gates homes and public buildings persisted. All of these characteristics played an important role in the development of the city into a thriving multicultural centre.

Jeddah and Hijazi food culture is one of the expressions of these facets, embracing the popular saying "Jeddah Ghair", which means "Jeddah is different". However, food traditions are very similar to the region's in general and to the national food culture. As a busy port city, Jeddah is a prime location for eating fresh seafood. Kabsa,[11] the national dish of Saudi Arabia, is actually considered the national dish for many Arab countries in the Persian Gulf and served in other countries, including Yemen, Somalia, Qatar, Oman, the United Arab Emirates, Bahrain and Kuwait. Many traditional regional dishes use similar ingredients, such as lamb, yogurt, rice, potatoes and dates. Islamic tradition and the nomadic tribal heritage, mentioned above, have influenced Saudi customs and table etiquette. Based on Bedouins' habits, daytime meals are usually smaller and the large meal is taken in the evening. Daily life is marked by the consumption of dates as snacks, and sweet tea, cola, *lassi* (yogurt drink) and buttermilk are popular beverages. The ritual of preparing the *Qahwa* (Arabic coffee) includes the use of four different pods to blend coffee grounds, spices, infused in hot water and served in small cups. Part of the ritual is the relationship between the guest and the host: to be polite the former should accept a cup, or even an odd number of cups of *Qahwa*.[12]

5.4.1 Exploring Jeddah: Food as the Sign of Epochal Changes and Connections

From a culinary standpoint, as shown by interviews and ethnographical fieldwork conducted with the students of the University of Business and Technology of Jeddah, the historic centre represents the place that most represents the cultural memory of food, history and traditions.

Historic Jeddah is a UNESCO (United Nations Educational, Scientific and Cultural Organization) world heritage site characterised by a unique architectural tradition, composed by tower houses built in the late nineteenth century by the city's mercantile elites and buildings influenced by the trade routes. The historic Al-Balad

[10] In the 1700s and 1800s, several riots characterised Jeddah, and the city was taken from the Ottoman Empire by the Nedji first but was later captured by Muhammad Ali Pasha, again under the rule of the Ottoman Empire.

[11] A mixed rice dish made of meat and vegetables.

[12] www.foodbycountry.com; www.tasteatlas.com.

(old town), near the Makkah Gate, is famous for its intricately designed houses built using coral from the depths of the Red Sea and boasting colourful *rawashan* balconies, known as *mashrabiyyahs*. However, Al-Balad, and the whole city, has a limited number of places in which a visitor or resident can find local food. In the old streets, it is possible to find a historical bakery (Fig. 5.1) that sells Arab bread and the traditional *shaboora*, which is usually eaten with savoury local cheese.

Nevertheless, among the old *souks* or markets in the historic area stands the most popular one, Souq Al-Nada, built more than 150 years ago, where it is possible to find fresh ingredients and some of the traditional dishes. In some places, it is possible to see the historical Jeddah through the collection of its old objects. The Hejaz House, Coffee-Museum (Fig. 5.2), a coffee shop and gallery inside an old traditional building, is famous for matching together food, drink and ancient objects.

Fig. 5.1 Historical bakery in Al-Balad

Fig. 5.2 Hejaz House (Al-Balad)

At another café-museum, the Layali Altarekheya coffee shop (Fig. 5.3), the traditional local food has been substituted by western food due to the low demand for traditional dishes and the high demand for different, particularly western foods such as doughnuts and muffins, from the local young people (Fig. 5.4). The experience of the traditional habits, conservation and display of the old furniture and objects seems to be in contradiction with the the taste experience of non-traditional food.

Even with the presence of different types of markets, such as Souq al-Alawi—the most traditional and extensive market in Jeddah, located around the mosques and *souks*—and popular shops and traditional crafts, the scene does not produce an image of typical foods and commensal habits and rituals. This is also due to the non-traditional menu options and an absence of tourist information.

Fig. 5.3 Layali Altarekheya coffee shop, Roof top (Al-Balad)

The lack of preservation of food as cultural heritage here is in contrast to the emphasis on heritage as displayed in outward facing media for non-Saudis. There exists a disconnect between the lack of traditional food products and dishes in these cafes, and the rhetoric of the institutional channels and media in supporting the concept of food typicality as a value to preserve and re-organise in the cultural memory of the territorial heritage. The web portal claims: "Plan your holiday in Saudi Arabia and explore its ancient heritage and culture, breath-taking outdoor activities and delicious local food" (www.visitsaudi.com). The portal constructs a nostalgic food heritage discourse, depicting positively traditional food identity, thereby contributing to the revalorisation of the cultural memory and control of the coherence of communication.

Fig. 5.4 Layali Altarekheya coffee shop menu

5.4.2 The Efficacy of the Discourse on Food Heritage

5.4.2.1 The Official Tourism Web Portal

In recent years, the national institutions supported the definition of local food through official channels, such as the Saudi Tourism web portal (now www.visitsaudi.com/en), to promote the country and the different regions and cities, focusing on a variety of sectors. It is important to point out that, between the first exploratory phase of the research in April 2019 and the following steps, the official Saudi Tourism website had been completely changed in terms of name, URL, and configuration. We will explore how the old version organised the discourse on food, and how the new web portal represents and reconfigures food identity that attempts to meet a need of innovation. The aim is to reflect on the quick change made to tackle the opening of the tourism sector in October 2019.

The old version of the website explained that there are diverse cuisines for all regions of KSA. In addition, it clarified that a number of restaurants contributed to the introduction of some Arab and Asian dishes to the Saudi table, which heightened the influences of different cultures on the Saudi food heritage. However, it emphasised that this is due to the presence of a large number of expatriates in KSA and the spread of many Eastern and Western restaurants. The food culture then is described and presented without mentioning any Saudi food ritual, but only displaying names, pictures and ingredients of the main dishes. The description only conveys that it is possible to taste traditional Saudi food in some of the restaurants located in the cities.

In contrast, the new web portal features and develops traditional and local Saudi Arabia foods and customs. The section "See and Do", for example, begins with the claim "Saudi Arabia's culinary scene is diverse, delicious and booming. From delectable gourmet cuisine to fragrant Arabic dishes, there's something to suit every appetite and budget", and is organised in three subcategories: "Regions", "Activities", and "Food and Drinks". By clicking on the third topic, the website opens other two groups: "What to eat" and "Guide". It is here that the user and consumer's expectations run high and ultimately clash with the dearth of information hinted at in the earlier page, providing a problem for the overall efficacy of the discourse.

In fact, the local and traditional food, which should be one of the most important source for destination attractions, contributing to the tourist experience, is misrepresented, and the message throughout the text-image relationship promotes dishes and beverages that do not really pertain to the country's food identity. This contributes to create an ambiguous meaning, where the pictures of a *cappuccino*, in the case of Riyadh, and of a waffle, in the case of Eastern Region, as well as the accompanying text, conflict with the user's expectations: "Café culture is ingrained in Arab communities across the globe, and Riyadh is no different", and "Wake up and smell the coffee, because the Eastern Province serves some of the very best breakfasts in Saudi Arabia. [...]".

5.4.2.2 The Saudi Press and the Role of the Institutions

A further examination of institutional official communication reveals similar food heritage frameworks. The 2016 website of the Saudi Commission for Tourism and National Heritage (SCTH), for example, published an article stating: "SCTH through sponsoring this event aims to introduce Saudi cuisine within the international food menu to be offered to the visitors and customers of restaurants and hotels of different categories across the Saudi cities" (www.mt.gov.sa/). The article also reported that a young Saudi chef, Annas Mubarak Al Saleem, at the venue said, "We are ready and very proud to present the Saudi food to the world through sophisticated hotels" (www.scth.gov.sa),[13] referring to Saudi kitchen's distinction and uniqueness and its ability to be developed to keep up with different tastes.

Moreover, regarding the events concerning food culture, the institutional website of The Saudi Center for International Communication (CIC)[14] promotes the preservation of the national heritage through the celebrations and events featuring products and activities, including folk food. Al Janadriyah is an annual festival sponsored by the Custodian of the Two Holy Mosques, is the most significant event that manifests the leadership's interest in "authentic" Arab heritage, culture, traditions and values. As stated in the article referred to above, "all regions of the Kingdom will be present at Al Janadriyah Festival to showcase their folklore, handicrafts, delicious folk foods and heritage through their wings, reflecting the kingdom's authentic urban heritage. They will receive visitors throughout the festival and through most beautiful performances and wonderful folk paintings" (arabnews.com).

Turning to the press, in an article dated 10 June 2018, *Arab News* discusses the traditional food during Ramadan, focusing on Jeddah and stating: "seasonal food stalls or kiosks, commonly known as *Bastas*, are among Jeddah's most popular features during Ramadan" (arabnews.com), with little information about food characteristics and practices. A few months later, another significant event related to the food heritage that was promoted in an article entitled "Jeddah Food Festival woos residents and visitors alike" did not contextualise the importance of the traditional food, but reported the words of Mazen Batterjee, Vice-President of the Jeddah Chamber:

> The beautiful atmosphere of Jeddah and the wonderful climate that it enjoys, gives more delight to the various pavilions of the festival, all of which are recreational means that meet all the needs of the Saudi family during school holidays and enhance the status of domestic tourism with Jeddah's great potential as an international shopping centre, a forum for all tastes and an attractive market for all local and international products. (live.saudigazette.com.sa)

The public website describes all the crucial concepts that define food identity, such as private home food experience, local resources, cultural influences and rituals, but not in depth. However, the public intervention plays a fundamental role in food distribution and consumption, since a law mandates that the number of calories for each dish must be displayed in every menu, changing in food choice trajectories, but also in food management, resources and standards now reconsidered.

---

[13] At the time of publication, the institutional URL has been changed, while the links to the related webpages have been removed.

[14] Established in August 2017, the CIC is an initiative of Ministry of Media to facilitate relationships with the global media community and as central source for the information in the country (cic.org.sa).

5.4.3 The Role of Religious Rituals in Remembering and Forgetting

As discussed above, the impermeable connection between food and religion exists not only in relation to the Halal concept but also with regard to holiday celebrations. Though not explicitly codified in the public and private channels, food heritage cuisine and food identity are mostly preserved in the rituals and religious life of people. *Ramadan* is a perfect example of this as its ritualised production and consumption, and ties to a unique setting, bind individuals across time and space through discourses of patrimony and inheritance. As with other forms of heritage, food invokes a sense of inherited tradition that must be preserved from the evanescence of time, particularly in the face of rapid transformation which threatens the integrity of the ethnic groups (Di Giovine 2014). To be passed on to future generations, therefore, heritage cuisine often defines itself against the new and globally diffuse, between conservatism, as social memory, and capitalism, new habits and traditional rites. Food rituals here appear to be the expression of identity through the respecting of values and norms associated with food consumption that influence the social practices that are reorganising everyday life. During Ramadan Saudis' food routine is reorganised, requiring fasting from sunrise to sunset. Thick soups, bean salads, *tabbouleh*, stuffed vegetables, flatbread, hummus, and rice compose the holiday's meal, while snacks and appetisers include dates, nuts and raisins. In the month-long celebration of Ramadan, the sunset meal, called *Iftaar*, assumes the aura of an experience in the restaurants, homes and streets of Jeddah. Roaming the streets during Ramadan is a particular experience for the autochthones as well as for the expatriates and visitors (Fig. 5.5).

By valorising the historical face of the city, and using the street as the scenario where Saudi heritage is displayed, food experiences become part of the cultural identity of public spaces. This experience that seems to be in contrast "with 'everyday' or basic eating, [and contains] a search for "authenticity" and "distinction" (Richards 2012). This is true, in particular in the case of non-Saudis looking for typicality and a total full immersive experience, in "stockpile of knowledge, traditions, memories and images" (Scott 2010) used to create idiosyncratic food consumption experiences. The many dishes offered as street food over the Ramadan are characterised by the diversity of their origins. Examples are *Sambosa*, snacks originally from India, *Balila*, chickpea soup, from Lebanon, *Mantu*, dumplings with a meat mixture, originally from China, and *Manaeesh*, the Arabian version of pizza: original dishes from central Asia and other neighbouring countries that have been adapted to the Saudi taste and are well integrated into the culture.[15]

Based on the different descriptions and the respondents' observations, we see that the taste, especially its variations based on some ingredients, is a key criterion for evaluating the authenticity of local Saudi dishes, though without recognising a real

[15] Other dishes served during Ramadan are *Yughmish*, a leavened bread, *Mutabaq*, a mouth-watering original from Yemen, *Luqaimat*, doughnut-like dumplings, *Sobia*, a drink, and the most traditional one for street food in Ramadan in Jeddah, *Kibdah* (fried liver).

Fig. 5.5 Ramadan celebration in Al-Balad (2019)

difference. The shape is relied upon to a lesser degree in determining authenticity. Moreover, food identity seems to be more a matter of private life, organising values and people habits.

Furthermore, it is interesting to see how some dishes considered traditional authentic Saudi food, as emerged from the interviews and informal discussions with native Saudi citizens, have been renamed to ensure these values. To quote Lotman, "in order to change from 'alien' (*chuzhoi*) to 'own' (*svoi*) this external culture must, as we can see, submit to a new name in the language of the 'internal' culture. The process of renaming does not take place without leaving a trace of that content which has received the new name" (2009: 133). The intersection with other cultural structures creates a condition of mutual tension, causing a distortion of their essence.

5.4.4 Adaptation, Authenticity and Identity Crisis

Through in-depth interviews and the focus group, we surveyed autochthones and expatriates to investigate both their consumption of traditional food and their conception of authentic food. Authenticity was an important issue while discussing their experience. Even though this term persisted as an imprecise concept in their reflection, it was clear that they attempted to associate it to the traditional food original from Saudi Arabia. Noting that the country has borrowed traditional dishes from other countries, the characteristics of Saudi food is identified by its abundance of the options, and the use of natural ingredients, such as olive oil. However, it is clear that the adjustments made due to the influence of other cultures, and the presence of many junk food options, are considered part of the erosion of authenticity. In particular, when informants were asked about the availability of traditional food in commercial venues, such as restaurants, and also asked where they go to eat traditional dishes in Jeddah, they stressed the fact that the food served in the restaurants is essentially local, but not necessarily traditional. When asked what exactly they mean by local food, informants mentioned that local food is not fully traditional. It was altered, they indicated, through the process of borrowing other countries' dishes, such as *sambosa*, or Yemenite food specialities, adjusting the ingredients, changing some dishes' names, making them acceptable and appropriate to be part of their culinary background (Avieli 2013). While these culinary intrusions and adjustments are effectively considered part of local rites and habits, these strategies, combined to the lack of a clear representation of "traditional food" through institutional and social channels, erodes and makes ambiguous the authenticity of the food experience.

Also contributing to shifting notions of local and authentic is that in the collective memory, the process of forgetting includes some dishes. One expatriate respondent who has lived in Jeddah for twenty years, mentioned *talbeena,* a soup made on a mixture of barley flour and water, to which milk and honey is added, considered a medicine by the Prophet Mohammed. Though the expatriate informant explained that the name is forgotten and the soup is no longer eaten, one of the autochthones noted that it is still known and still part of Saudi families' food habits. Two other examples of purportedly traditional dishes whose origins are fading from the collective memory are *Kabsa*, which seems to belong to different Arab countries, and *Muttabak*, that, as stated from one of the interviewees, "originally came from Yemen but now it is becoming Saudi Arabia dish because everyone is eating it". Given that Jeddah is a historical crossroads of many countries and cultures, perhaps the best representation of its cultural identity is one of diversity, perhaps even characterised as an identity crisis, both for its inhabitants and also those looking in from the outside.

This confrontation between Saudi Food and other Arab countries' food traditions reverberates across social and institutional contexts and reveals an unpredicted outcome. By associating the food heritage with rituals, such as Ramadan, and at the same time with other cuisines and fast food, the local food becomes a marker of epochal transformations. However, the opening of the city to globalisation heightens the felt need to reinvent traditions in some restaurants. Working against the international influences paradoxically adds value to their traditional food items and dishes even as they continue to absorb and be altered by other cuisines. Moreover, the traditional restaurants rarely present a strictly local menu; it is always accompanied by other cuisines such as Yemeni, Jordanian, Indian and Lebanese cuisines. Adding an aesthetic value to the local dishes is the impact of such global "hierarchies of value" (Herzfeld 2004) on food heritage, in relation to the cuisines from neighbouring countries, which sometimes dilutes the purpose and strategy for attracting consumers. If the regional or national cuisines are symbolic representations of the nation and of the identity of a people (Appadurai 1988), diversity is described here as an essential resource necessary to the creation of Saudi food as it is present constituted. That heterogeneity, as a foundation of the semiosphere, represents the component of social identity (Régnier et al. 2006). In this case, the principle of territoriality is the criterion to identify Saudi food as traditional food. This highlights that position, discussed by Paul Ricoeur in *La condition d'étranger* (2006), at the foundation of the opposition foreign/member that seems to serve as a basis for the existence itself of the Saudi food identity.

The use of the projective techniques, where the respondents were asked to pick the pictures of two dishes among a collection of diverse Saudi meals, to state if they represented authentic Saudi food, and share a reflection about them, regarding experience and memory of the dish, reveal different consideration of the authenticity of Saudi food. Both the autochthones and expats highlighted that local food cannot be considered as traditional, and that the main difference lies in the strong influences that affected the Saudi food identity. Yet a different outcome was discovered during the focus group. As part of the exploratory research design, it was conducted to explore autochthones' experience and perception of the topic of interest and to better understand the complexity of the phenomenon. The sample depended on the stratifications of the population of interest by considering the demographic aspects such as. The convenience sampling technique guided the recruitment process identifying a first sample of six UBT students, among male and female, in the age of 23–25 years old. Data collection and analysis were followed as an interactive process until the point of saturation was reached.

The emphasis was placed on the interaction between participants, in particular in the light of strong dynamics activated by the different proveniences, based on the origins of the respondents' families, in terms of Saudi regions and tribes. For example, some of the participants stated that the *sambosa* is an original snack from Saudi Arabia, ignoring its Indian provenance, while other participants highlighted that the traditional dishes are local but not originals from the Hijab region.

5.5 Conclusion

This study has revealed an interesting finding that both autochthones and expatriates shared the same point of view regarding the confusion around the idea of authenticity of the Saudi food. Even indicating a different level of knowledge regarding the traditional dishes and their origins, both believe that food heritage is dealing with an identity crisis of sorts. This brings us back to the problem of translation as one of the several elements on what to pay attention when it comes to the consideration of the "interpretation of food as a language and the analysis of the processes of translation between different foodspheres" (Stano 2016: 20). In the current study about the Saudi foodsphere, this would generate from a blend of cultures that, beyond the food heritage of Bedouins and from the Islamic history of Saudi Arabia, created a new local food identity, which is now subjected to a forgetting process. We refer with this to the areas of valorisation of national collective memory as a form of resistance to forgetfulness: presenting local food not exactly as local, not sufficiently highlighting the food traditions, is only one of the ways in which the country's gastronomic heritage is manipulated, creating a hidden side for new generations, expatriates and tourists. The food identity becomes the expression of that *hybridity* and of the connection between the search for an identity and the development of awareness concerning the recognition of the other as integral part of that system (Lévi-Strauss 1977). Stating that Saudi food is not Saudi here means recognising its real identity. It reveals itself absorbing the others in a composition of histories and travels impacting, renegotiating and creating local cultures and consumer tastes, shifting from an *omnivorous* to an *inclusivist* consumption. Food identity, and the idea of an *identityin progress*, now, more than ever, must be considered by taking into account the fact that Saudi Arabia is opening for the first time to tourists, implying that this could be the case of epochal transformations, not limited to the practices for travelling and exploring another country, region or city, but inclusive change which is impacting on socio-cultural resistances, such as gender equality, women empowerment and human rights. At the same time, food heritage expresses the common concern for historical continuity and protection of a shared sense of community and our affiliation to a collectivity as members (cf. Boero and Greco 2018).

References

Al-Rasheed, Madawi. 2002. *A History of Saudi Arabia*. Cambridge: University Press.

Appadurai, Arjun. 1988. How to make a national cuisine: Cookbooks in the contemporary India. *Comparative Studies in Society and History* 30 (1): 3–24.

Avieli, Nir. 2013. What is 'Local Food?' Dynamic culinary heritage in the World Heritage Site of Hoi An. *Vietnam. Journal of Heritage Tourism* 8 (2–3): 120–132.

Barthes, Roland. 1975. Reading Brillat Savarin. In *The Rustle of Language*, 250–270. Berkeley and Los Angeles: University of California Press.

Becuţ, Anda Georgiana, and Kàtia Lurbe i Puerto. 2017. Introduction. Food history and identity: Food and eating practices as elements of cultural heritage, Identity and social creativity. *International Review of Social Research* 7(1): 1–4.
Boero, Marianna, and Cristina Greco. 2018. Suspended identities. The concept of Ius soli among memory, boundary and otherness. *RIFL/SFL*: 16–25.
Brillat-Savarin, Jean Anthelme. 1838. *La physiologie du goût*. Paris: Editions Charpentier.
Di Giovine, Michael A. 2014. The Everyday as Extraordinary: Revitalization, Religion, and the Elevation of Cucina Casareccia to Heritage Cuisine in Pietrelcina, Italy. In *Edible Identities: Food as Cultural Heritage*, ed. Ronda L. Brulotte, Michael, A. Di Giovine, 78–92. London and New York: Ashgate.
Handler, Richard, and Jocelyn Linnekin. 1984. Tradition, Genuine or Spurious. *The Journal of American Folklore* 97 (385): 273–290.
Herzfeld, Michael. 2004. *The Body Impolitic. Artisans and Artifice in the Global Hierarchy of Value.* Chicago and London: Chicago University Press.
Lévi-Strauss, Claude. 1977. L'identité, Séminaire interdisciplinaire dirigé par Claude Lévi-Strauss, professeur au Collège de France, 1974-1975. Paris: PUF.
Lotman, Jurij M. 1984. O semiosfere. *Trudy po znakovym sistemam* 17: 5–23. English edition: Lotman, Jurij M. 2005. On the Semiosphere (trans. Clark, Wilma). *Sign System Studies* 33: 205–229.
Lotman, Jurij M. 1992. *Kul'tura i vzryv*. Moscow: Gnozis. English edition: Lotman, Jurij M. 2009. *Culture and Explosion* (trans. Clark, Wilma). Berlin: Mouton de Gruyer.
Matta, Raúl. 2013. Valuing Native Eating: The Modern Roots of Peruvian Food Heritage. *Anthropology of Food* S8. https://journals.openedition.org/aof/7361. Accessed 27 March 2020.
Md Ramli, Adilah, Mohd Salehuddin Mohd Zahari, Mohd Zulhilmi Suhaimi, and Salim Abdul Talib. 2016. Determinants of Food Heritage Towards Food Identity. *Environment-Behaviour Proceedings Journal* 1(1): 207–216.
Régnier, Faustine, Anne Lhuissier, and Severine Gojard. 2006. *Sociologie de l'alimentation*. Paris: La Découverte.
Richards, Greg. 2012. Food and the Tourism Experience: Major Findings and Policy Orientations. In *Food and the Tourism Experience*, ed. Diane Dodd, 13–46. Paris: OECD.
Ricoeur, Paul. 2006. La condition d'étranger. *Esprit* 3: 264–275.
Scott, Allen J. 2010. Cultural economy and the creative field of the city. *Geografiska Annaler: Series B Human Geography* 92 (2): 115–130.
Stano, Simona (ed.). 2015. *Cibo e identità culturale/Food and Cultural Identity. Lexia* 19–20. Rome: Aracne.
Stano, Simona. 2016. Introduction: Semiotics of Food. *Semiotica* 211 (1/4): 19–26.

Websites

Arab News Website. https://www.arabnews.com/.
Center for International Communication (CIC). https://cic.org.sa/.
Food and Agriculture Organization of United Nations (FAO). https://www.fao.org/.
Saudi Commission for Tourism and National Heritage (SCTH). https://scth.gov.sa/.
Saudi Gazette Website. https://live.saudigazette.com.sa/.
Saudi Tourism web portal. https://www.visitsaudi.com/.

Cristina Greco is Assistant Professor in Communication, and Vice Dean at the Jeddah College of Advertising, and Head of the Communication Research Unit and Research Operation Office at the University of Business and Technology (KSA). Her research interests include cultural semiotics, visual culture, urban semiotics and emerging media with a focus on the theories of collective memory. She was Visiting Professor at the Polytechnic University of Valencia, and also the University of Gent. Her publications include "Suspended Identity: The Concept of Ius Soli among Memory, Boundary and Otherness" with Boero, M. (*RIFL/SFL* 2018: 16–25); "Dall'autobiografia al documento. Il graphic novel tra memoria archiviata e svelamento dell'illusione in Anne Frank. The Anne Frank House Authorized Graphic Biography" (*Signa*, 26, 2017); and "Translating Cultural Identities, Permeating Boundaries. Autobiographical and Testimonial Narratives of Second-Generation Immigrant Women", in Olivito E., ed. (Routledge 2015).

Chapter 6
Bittersweet Home: The Sweets Craft in the Urban Life of Tripoli, Lebanon

Henry Peck

Abstract The city of Tripoli, Lebanon, is often disparaged for its social issues but praised for its formidable Middle Eastern sweets. My research examines why Tripoli became renowned for sweets, and why this craft endured in the face of urban change. I briefly outline the features that facilitated the production and trade of sweet goods for many centuries, suggesting that their subsequent commodification at the dawn of modernity helped cement their place in the city's identity. Today sweets are arguably the city's most successful product—a point of pride for many residents, but of frustration to some who see this as indicative of failings in heavy industry, or a shallow representation of their city as a whole. Following from Barthes' division of the nutritional and protocol values of food, I find that while Tripoli's sweet pastries provide a caloric source of varying need to consumers, they have singular importance in social and religious occasions, signifying status, hospitality, and in-group identity, among other dynamic meanings. Their shifting meanings also reflect changing tastes and health narratives, as seen in hybrid and "diet" varieties, emblematic of the modern period.

Keywords Foodways · Confectionary · Heritage · Craft · Modernity

6.1 Introduction

This study responds to the questions of why the Lebanese city of Tripoli is renowned for sweets and why this renown has endured in the face of urban change. The contemporary production of Middle Eastern sweets does not require the kind of natural environment or agricultural proximity that some place-linked goods, such as wine or maple syrup, call for.[1] It takes place in cities throughout the Middle East and beyond,

[1] Even with these items the cultural domain appears to be more significant than the environment in creating the *goût du terroir*. See Trubek (2008: 26).

H. Peck (✉)
Faculty of Oriental Studies, University of Oxford, Oxford, UK

S. Stano and A. Bentley (eds.), *Food for Thought*, Numanities - Arts and Humanities in Progress 19, https://doi.org/10.1007/978-3-030-81115-0_6

but Tripoli is one of few distinguished for the range and quality of its sweets, a whole type of which are not readily made elsewhere. This distinction has grown while the city's other crafts and indicators have declined.

I suggest sweets are ingrained in the city's identity through a combination of historical links, prevailing social and religious usages, and successful commodification. Tripoli's history is tied to maritime and inland trade, and sweet goods were among its major exports for centuries, thanks to the local availability of ingredients. When sugar production shifted away from the area, Tripoli continued to produce sweet goods, indicating their continuing local and commercial appeal. Yet the nature of their association with Tripoli also highlights the city's malaise; that sweets are its primary reference of success frustrates some residents who see it as a reminder of the decline of heavy industry or absence of grander symbols. Examining Tripoli's sweets craft offers an oblique entrance to life in the city and its issues, shifting habits and tastes, and the tensions of modernity.

6.1.1 What Are Sweets?

I use "sweets" as the general translation for *ḥalawīyāt*, the Arabic term for sweet pastries, desserts, candies, and confectioneries. These are, of course, types of food, and like all foods have not only a nutritional definition but symbolic meanings shaped by their particular usages (think of the expression "as American as apple pie"). As Roland Barthes noted in 1961, "to the scholar, the subject of food connotes triviality or guilt" (Engl. Trans. 2008: 28), but its relation to certain "protocols" is packed with information. I suggest this is particularly acute with sweets in Tripoli because their social, cultural, and psychological dimensions supersede their nutritional value—for a start, they are not eaten during mealtimes, and they constitute a vital element in religious holidays. As such, it is helpful to think of sweets as a system of communication.

6.1.2 Historical Overview

Over its long history Tripoli grew from Phoenician trading settlement to increasingly prominent coastal city under Mamluk rule, to fractious site for the Ottoman authorities and host to further rebellious activity in the twentieth century. Fertile land and the port supported the production and trade of sweet goods and fruits, and later silk and soap. Competition with Beirut intensified in the nineteenth and early twentieth centuries, but Beirut grew at a much faster rate and was established as the dominant city in the newly created Greater Lebanon (1920), where the French colonial power largely overlooked Tripoli. In the three decades following Lebanese independence in 1943 Tripoli experienced cosmopolitanism and industrial decline, becoming home to many outsider groups and pan-Arab currents (Larkin and Midha 2016: 182). But the civil war (1975–1990) saw middle class flight, a period of Islamist control, and the beginning of nearly four decades of Syrian occupation. It has left the city prone to fresh outbursts of violence in the years since. In 2015 the UN estimated that 57% of the population in Tripoli is poor or deprived (UN ESCWA 2015).

6.1.3 Hallab

Central to this study is the name Hallab, used by the two largest sweets companies in Tripoli today. Like almost all of Tripoli's sweets companies, Abdul Rahman Hallab (also labelled "Hallab 1881") and Rafaat Hallab (also labelled "Rafaat Hallab 1881") are family-run businesses that split from a parent company. Both claim to have been established in the year 1881, but Abdul Rahman Hallab has been a distinct operation since the 1970s, and Rafaat Hallab since the 1940s. The family name is important to the mythology around the product, even as their businesses outgrow familial structures. The different priorities the two businesses have placed on marketing and product is revealing of the importance of advertising in advanced capitalism. Both businesses have made crucial contributions to Tripoli's enduring renown for sweets.

6.1.4 Methodology

In my data collection for this study I used a combination of interviews, ambient recordings, ethnographic field notes, and pre-existing recordings. A third of the interviewees were involved in the sweets business in Tripoli; the remainder were long-term Tripoli residents (a few no longer lived in the city but continued to work there or visit frequently). Of those not involved in the sweets business, the majority considered themselves active participants in city life and were educated up to a bachelor's degree or higher. The interviewees ranged in age from approximately 30–80, and I estimate that half were aged between 40 and 65. They were predominantly male, and the sweetshops were almost all staffed by men. Over the course of my visits, I also had many informal conversations with other people in Tripoli, with members of the Hallab family not involved in the business, and with several employees at various sweetshops. The interviews and conversations took place in English or Lebanese Arabic (and occasionally Modern Standard Arabic). The analysis presented in this chapter is abridged from a larger body of unpublished research.

6.2 City and Sweets in Time and Space

The food of Tripoli bears traces of different ruling powers. The material influence of the Ottoman Empire on the local sweets craft is clear, with the Turkish history of filo and baklava well-documented (Davidson 2014: 307). Baklava remains the predominant sweet pastry in both Istanbul and Tripoli, but the Tripolitan producers I spoke with were quick to distinguish their work from that of their Turkish counterparts, telling me Turkish baklava relies on excessive sugar proportions of 50–60%, as opposed to their own more subtle ratio of 20 percent (also somewhat reflective of health discourses, as examined below).[2]

The 23 years of French mandate rule would also leave a mark on Lebanese cuisine, though less clearly in Tripoli. The cookbook author Anissa Helou, who left Lebanon at the age of 21 and writes for a largely Western readership, describes this as the French "refining influence", "which probably explains why Lebanese food is that much more varied and refined than that of its Middle Eastern neighbours" (Helou 1994: 8–9). Beyond the implicit value judgment, categorising food on national lines assumes large cultural continuities. This is a common tendency among both nationalist ideologues and food writers (Holtzman 2006: 368). Instead, it is more useful to look at food variations within localised contexts. In Tripoli, the essential public cuisine (found in the plurality of longstanding restaurants and street carts) revolves around breakfast or lunch dishes of *fūl mudammas*, hummus, *sfīḥah*, and *fatteh*. These more closely resemble Damascene fare than that of Beirut. Savoury *kaʿak* (a flat bread) is available all day, and evening meals often feature fish along with mezze staples. *Manʾūsheh*, shawarma, hamburger, and sandwich spots are also common. Tripolitan writer and local historian Safouh Mnajjed has described the enthusiasm in the 1950s–1970s for European cuisine (Mnajjed 2013).

The sweets sector adopted French goods (such as macarons, croissants, and cakes), but was recognised elsewhere in the region in the twentieth century for the production of Middle Eastern sweets. The grandfather of the current directors of both Hallab companies took part in sweets fairs in Jerusalem in 1934, Beirut in 1935, and Damascus in 1936, where he was awarded a golden medal and certificates. When King Faysal of Iraq (whose removal by the French the Sunni Tripolitans had so resented in 1920) visited Tripoli, the city's sweets were of sufficient renown that Rafaat Hallab was asked to design a piece of confectionery to mark his visit, and the *Faysaliyya* was born.[3]

It is worth noting that by the middle of the twentieth century the indigenous ingredients of sweets production had largely disappeared from Tripoli, without diminishing commercial production. This also reflects the changing nature of the local economy away from small business towards modern formulae for quick profit. Sugar, citrus fruits, and blossoms remained important ingredients to the sweets business, but the city now imported sugar, and its famous citrus orchards that had constituted the city's major crop for at least a thousand years were dwindling, their landowners

---

[2] Refaat Hallab, interview.

[3] Refaat Hallab, interview.

selling or developing them in the increasingly profitable real estate trade (Gulick 1967: 95–97). The rapid development of land would become a defining business preoccupation in Tripoli and across Lebanon, and is emblematic of a broader trend away from skilled artisanal work to construction labour.

The place of sweets had long become a given—concurring with Sidney Mintz's explanation of how sugar became a commonplace necessity of modern life (1986: xxix). Modernity brought the transport, technology, and migration that enables the dissociation of food production from its environment, and yet Tripolitan sweets persisted. I argue this is because it was a product ripe for commodification under conditions ready for capitalist modernity. What began as a craft responsive to its environment was transforming into an industry, anchored by local affinities in production and consumption.

6.3 Sweets in Society

My analysis of sweets in society rests on a central idea of foodways as a form of communication, in which the production and consumption of food conveys information, carries and contains meanings, and underscores human expression in a given environment. Engaging with this topic inevitably brings up the aphorism of Jean Anthelme Brillat-Savarin to describe European culture, "Tell me what you eat and I will tell you what you are" (quoted in Terrio 2000: 237). Rather than adopting the descriptive implication of his line, I seek to align more closely with the approach taken by Casey Man Kong Lum and Marc de Ferrière Le Vayer, who extend the meaning from, at its simplest, "you are what you eat", to, "we are how we eat" (2016: 6). With this paradigm in mind, I try to show how sweets and their foodways reveal and inform social relations and aspects of urban life in Tripoli, including as metaphors for identity.

I suggest that sweets have a moral valence, conferred through ritual usage, hospitality, language, and the upstanding reputation of the confectioners. They may be consumed to present public expressions of status, class, or faith, and even for political purposes. Their diverse meanings stem from usage in both public and private spaces. These uses may become binding moments in which group identity and distinctions are reinforced. The memory of such moments in childhood is often an evocative symbol of time and place for older generations. As tastes change across generations in Tripoli, these values are modified and reformulated through new social occasions, with certain signifiers preserved. Simultaneously, Tripoli's renown for sweets serves to substitute for a lack of political and economic success. For a place that often looks to its past wistfully, sweets provide a compelling bridge to a more prosperous time.

6.3.1 Nutrition and Protocol

A young Tripolitan man remarked to me that, "Winter is cold, so we need sweets. Baklava gives fatty energy (*shahīm*), ghee in sweets gives energy.... Sweets means enjoying life. Honey equals happiness".[4] His statement, describing first a circumstantial utility and then an emotive power, introduces the divergent definitions of the edible object. It approaches Roland Barthes' description of food as having "a twofold value, being nutritional as well as protocol, and its value as protocol becomes increasingly more important as soon as the basic needs are satisfied" (1961; Engl. Trans. 2008: 34). The distinction is useful in thinking about the place of sweets in Tripoli, where these baseline values give way to polysemic associations for social and religious life.

As a sugary food, sweets do make up an important calorie source, particularly in low-income diets, but with the exception of breakfast *kanafeh* do not form part of regular meals (where dessert consists of fruit, if anything). As symbolic units, they often give weight to the occasion in which they are served. Mintz documents that over time,

> sugar products became even more important to the poor than they once had been for the wealthy—as sources of calories even more than of status—and since the occasions for eating them multiplied, new uses and meanings arose at a great remove from the practices of the privileged. (1986: 152)

This is apt for Tripoli, with its high level of poverty and multitude of settings for sweets consumption across socioeconomic levels. Continuing in this vein, no food has only nutritional value: in eating we also "consume socially produced meanings" (Heinz and Lee 1998, quoted in Leeds-Hurwitz 2016: 189). In Tripoli, confectionery eaten between meals is linked to rituals including public social engagements, acts of hospitality, or the marking of life milestones. A tray or box of sweets is a visible fixture in most living rooms, ready to be offered to guests, suggestive of a hospitable home. These practices are longstanding but also dynamic, reconfigured as tastes and generations change.

The occasion of the *visit* showcases the values sweets have in both quotidian and special events. In fieldwork visits to homes or offices around Tripoli, I was invariably received with tea or coffee and sweets. I remarked on this to several of my interlocutors, who responded that it is considered standard practice in a spontaneous or planned house call, as well as a feature of regular occasions such as women's morning gatherings (*ṣubḥiyāt*) or afternoon teatime.[5] While it was taken for granted as proper hospitality to offer sweets to guests, they acknowledged it is equally customary for visiting guests to present a gift to their hosts, with the secure choice in Tripoli being a tray of confections from a sweetshop. This extends to special occasions. The same interlocutor described sweets as the appropriate gift for three very different types

[4] Mohamad, conversation.

[5] Samer Hallab, interview; Mohamad, conversation.

of formal occasions: joys and celebrations, sorrows and deaths, and religious holidays (*ʾafraḥ, al-ḥuzn,ʿeīd*). Within the first category, he gave the example of the preparation and gifting of particular sweets for specific occasions, such as *maghlī* to commemorate the birth of a baby, and *snīnīyye* (a diminutive playing on the word for tooth) on the appearance of a child's first tooth.[6] Religious holidays, such as the month of Ramadan where the daylight fast is broken every evening by the iftar meal, are described as presenting not only a tradition of having a table full of sweets, but creating a bodily "need" for them.[7]

We can read from these examples the importance of sweets to a range of social settings, that in turn reinforce their staying power. Sweets become synonymous with certain occasions, their material presence and consumption requisite elements of the event. They convey particular information within specific contexts, following from Barthes' idea that such "food serves as a sign not only for themes, but also for situations; and this, all told, means for a way of life that is emphasised, much more than expressed, by it" (1961; Engl. Trans. 2008: 33). This idea of emphasis over expression is significant, as they do not on their own denote a lifestyle or situation, but are a key component in the construction of situations. I found these situations could be leisure, business, celebration, or activity—just as other foods correspond with and emphasise the practice of sports (energy bars, protein drinks), different types of work (a breakfast of *fūl mudammas*, packaged sandwiches, "business lunch" specials), weddings (wedding cake) and so on.

Similarly, the custom of confectionery exchange is revealing of social dynamics. The apparently voluntary gift of sweets is in fact one part of an ongoing reciprocal exchange, "given, accepted, and repaid under obligation" (Marcel Mauss 1925, referenced in Terrio 2000: 256). As these gifts flow in both directions, they can help to level power differences in relationships that are not symmetrical. Conversely, the extravagance or expense of the gift may vary, which can entrench class positions and engender social prestige. Susan Terrio has documented this in the exchange of chocolate—at one level a cheap postmodern commodity, at another, an expensive honorific food—among the French middle classes. She finds it has become a strategic component in asserting social distinction, which links to Bourdieu's claims of taste and eating styles being deeply embodied markers of class difference in the 1970s (Terrio 2000: 258–259; cf. Klein 2014: 9). I did not witness a sufficient number of confectionery exchanges in Tripoli for direct analysis in that setting, but my discussions and observations of producers suggested comparable scales of distinction and taste, so that different products when gifted can emphasise social status, wealth, generosity, kinship, or other meanings.

Sweets traverse both public and private realms in Tripoli, adding to the ways their social meanings are generated and reproduced. An architect and lifelong Tripolitan expressed loyalty to her neighbourhood store in Abu Samra, but this did not prevent her from patronising stores elsewhere in the city, or from professing the superiority of certain stores for particular items. Opinions on the latter were shared among my

[6] Mohamad, conversation.

[7] Mohamad, conversation; CH, interview.

interlocutors, for example all stated that Nouh Haddad does the best *ḥalawat al-shmayseh*—it makes only this and *jazarieh*, a kind of preserve made from carrots and often mixed with nuts. That these preferences were held in common suggests a similar judgment of quality as well as a shared sense of city identity as it relates to insider knowledge.

6.3.1.1 Signalling Taste

The gifting of a box from any store signifies moderate expense, whereas the price of a box from Abdul Rahman Hallab can raise eyebrows, and this meaning is broadcast in punchy packaging—reinforced bags in the company's trademark green colour with the company's palace logo, itself a symbol of grandeur, prominently featured on the side. Its success in marketing this image of prestige confirms the privileged status it retains even among those who prefer the sweets of other stores. It gives the impression of luxury, and its notable expense links its public consumption to an aspirational lifestyle.

Just as carrying an Abdul Rahman Hallab bag becomes a visible signal of taste, so does sitting in the store's dining area and hall. Eating in a sweetshop presents other considerations to eating at home or in another private space, conveying a public manifestation of status (Ferguson 2014: 237). This is reflected too in the presentation of sweets in these spaces, where looking precedes eating, and must impress accordingly. As Deeb and Harb (2013) point out in their study of expressions of leisure in Beirut, "taste is also about more than a café's décor or menu, though those elements do play a role" (209). Rather, taste "both marks class and produces class boundaries.... How people consume—including how they consume leisure—indicates and produces social hierarchies" (Ibid.). Therefore, the high prices and luxury treatment at Abdul Rahman Hallab generates an appeal beyond the quality of its products. The public display of such consumption is a spectacle, which for some is socially rewarding, where ostentatiousness is an indicator of status. Anthropologist Christa Salamandra observed similar phenomena in Damascus, where she links public consumption of repackaged "old" tastes to modern identity construction among the urban elite (Salamandra 2004: 2–4). In Lebanon, luxury ideas of prestige figure into many other expressions of consumption, cars being the obvious example (Cochrane 2019). And indeed in Tripoli, Hallab sits astride old and new types of café consumption, with a strip of new cafes in the upmarket Dam & Farez area even more popular among the city's wealthy and on most evenings fronted by a row of expensive cars.

The implicit signalling in eating choices has also been long evident in the option of the patisserie, as seen to be denoting European tastes. Today, the larger sweetshops have their own patisserie offerings (such as cakes and macarons), but Rafaat Hallab segregates those items to its second branch, and Haddad has a separate storefront next door to its sweetshop. The distinction at Haddad is heightened by the contrast in signage: quite apart from using different logos, the name and description of the sweetshop is written in Arabic, while for the patisserie it is in French. These hybrid stores offer a diversification of lifestyles in culinary taste, complicating class difference rather than eliminating it (Klein 2014: 9).

6.3.1.2 Religion

Overlapping with their place in social life is the position sweets hold in religious life in Tripoli. Here sweets have further symbolic meanings, from the ritual element in religious festivals, to permitted public indulgence. They feature in the religious festivals of Tripoli's Christians and Muslims, but are particularly prominent for the latter. It is worth noting that the two best known centres of sweets production in Lebanon, Tripoli and Sidon, are overwhelmingly Sunni, as are their confectioners. Cookbook author Anissa Helou speculates that this stems from the difference between the Christian austerity throughout the 40 days of Lent and the nightly feasting during Ramadan in which sweets play a central part, as well as in the holiday feast celebrating its ending, Eid al-Fitr, and the even larger holiday feast about two months later, Eid al-Adha (Helou and Pugliese 2015: 6).

On an evening during Eid al-Adha in August 2018, two sweetshops I visited still had some *maᶜamoul* in trays but were unable to sell me any because their entire inventories had been bought out in advance. These stores were full of customers being served by salesmen in distinctive uniforms, assembling large orders amid excited chatter. Certain sweets are associated with each festival, and do a roaring trade at these times, but in Tripoli's sweetshops almost all of these are now available at any time. Samer Hallab, the CEO of Abdul Rahman Hallab, confirmed that "we used to make special items in Ramadan. Now we make them all year round".[8] This parallels the sale of hot cross buns in the UK, which originated as a food to mark the end of Lent, but are now also sold throughout the year, and have likewise incorporated variations (such as chocolate) for changing tastes. In both cases, general commercial interests move beyond the ritual religious significance of these items. Yet their enduring ties to religious holidays reflect powerful associations of identity and community.

The consumption of sweets entwined with a religious occasion can be a potent symbol of participation in the faith or group. As Mintz and Bois note, "In consecrated contexts, food 'binds' people to their faiths through 'powerful links between food and memory'" (2002: 107). Simultaneously, it can reinforce religious boundaries that might exclude others. Emiko Ohnuki-Tierney suggests that part of its power derives from being an ingestible metaphor for a social group:

> When each member of the social group consumes the food, it becomes a part of his or her body. Thus this important food becomes *embodied* in each individual and functions as a *metonym* by being part of the self. In addition, when individual members of the social group eat food together, they are participating in a communal action that bonds them as a group. (1999: 244)

Most of my experiences of Eid al-Adha and Ramadan iftars in Tripoli took place in public spaces, where there were no obvious barriers to entry for non-observers, but

[8] Samer Hallab, interview.

the consumption of sweets at these occasions remains linked to an idea of a particular faith (Mintz and Du Bois 2002: 107). Similarly, at a Christmas market at the city's international fair complex in December 2018, I encountered both Christians and Muslims, and secular symbols of the holiday (tinsel, snowflakes, Christmas trees, but no nativity scene), suggesting secular commercialism over manifestation of faith. The Tom Sweets company had a stall alongside various other food vendors, but there is not (yet) any specific food tied to the occasion of the Christmas market in Tripoli.

At the same time, it would be short-sighted to suggest that the consumption of sweets in a ritual tied to faith equates to having a strong religious identity. Anthropologist Jon Holtzman questions the extent to which we should read identity through food, "particularly when such an identity may not have much life outside festivals or public displays" (2006: 366). The enduring popularity of sweets in such occasions does invite public circulation and exchange, even if some of the festival sweets are made at home. Conversely, they may be overlooked for much of the year until the festivals, when sales can quadruple (Armstrong 2012). This difference is marginal compared to the Gulf market according to Refaat Hallab (a director of Rafaat Hallab), who complained of sweets only being popular in Riyadh during Ramadan, whereas you could sell cakes "everyday".[9] Outside of religious festivals, sweets offer a licit sensuous indulgence. Shops are open late into the evening, and even all night during Ramadan, generating social activity. Helou suggests this "this love of sweet things is a result of the Muslim prohibition against alcohol, with sugar from sweets replacing the sugar derived from alcohol" (Helou and Pugliese 2015: 5).[10] But as described, the reasons for the city's sweet tooth are more nuanced, and their meanings are far from fixed.

6.4 Product as Palimpsest

There may not be all that much that distinguishes the baklava at Abdul Rahman Hallab from that at Rafaat Hallab, but myriad other factors inform the experience of consumption and shape tastes and allegiances towards one or the other, even as the products themselves move towards homogeneity. Changing habits in sweets consumption may have more to do with shifting ideas about nutrition than economic factors, as Barthes' analysis of eating trends suggests (1961; Engl. Trans. 2008: 29). Sugar was first considered a remedy, constituting a common element in Arab medicines, and was also a purely luxury product until the late eighteenth century (Heine 2018: 104). Showing how far these meanings have changed, today luxury tastes privilege the experience of the high-end sweetshop more than the sweets themselves—while, paradoxically, consuming the most "humble" of local foods can confer social distinction (Klein 2014: 17). This was evidenced at the Christmas

[9] Refaat Hallab, interview.

[10] Alcohol is actually derived from the breakdown of sugar, though alcoholic drinks contain residual amounts.

market in Tripoli (where the entrance fee was 5000 lira, just over $3, disposable income that many do not possess) in the presentation and popularity of the *mujaddara*, an inexpensive but ancient dish of lentils and rice. At the fair it was billed as traditional and a novelty and sold at a handsome price.

6.4.1 Innovation and Health

While both Hallab companies and the major stores continue to offer their staple products, they showcase a widening array of new creations. Refaat told me that when his great-grandfather started the store it made perhaps ten kinds of sweets; today the store sells 70–80 different varieties. This is partly a response to the evolving tastes of younger generations. A senior employee at Abdul Rahman Hallab confirmed the company is concerned the next generation does not have the same appetite for Middle Eastern sweets, and is trying to win it over with hybrid products. She told me, "We are proud of [traditions], but at the same time we work on supplying what the market needs…. That's why [we] invented the chocolate *kanafeh*…. we want children to like *kanafeh*, now they like it". Chocolate *kanafeh* might seem unthinkable to another generation, but many of the shops have incorporated Western cakes and chocolate into their product lines, reflecting demand for different and sometimes blunted flavours, along with reinvention.

This ceaseless desire for novelty is another modern phenomenon. Historian Lynne Hunt critiques the modern emphasis on "restlessly iterating the new" and its excessive concern for innovation, which is evident in the Hallab stores (Hunt 2008: 82). Perhaps she undervalues the sense of possibility it allows, however. Some of the new concoctions verge on the bizarre—Refaat shared with me his latest line, sweet sushi (made entirely of sweet materials)—but they may allow their confectioners creative agency and imagination, which is arguably preferable to standardisation's logical end, the infinite mass production of the same object. As Sami Zubeida explains in reference to the changing meaning of mezze, which was originally an accompaniment to alcohol rather than a meal in itself, "current culinary discourse and practice are on constant lookout for innovation and invention, which is often buttressed by an invented tradition" (Zubaida 2014: 222). While food never stays entirely static, the deliberate and constant attempts at innovation are characteristic of our modern period.

Similarly, international discourse around health is shaping new products. These include sweet syrup made with a sugar substitute, and gluten-free and "diet" lines of popular items. They endow the sweets the new meaning of eating within a local tradition but with modern health awareness. Or, to use another food idiom, they appear to give customers the power to have one's cake and eat it—consuming rich heritage foods while keeping a clear health conscience. A young and mobile life coach remarked how Lebanese people are returning from abroad to try to "break the model", as with "TAQA", a new wholesale bakery on the edge of Tripoli producing preservative- and GMO-free "health snacks" in the form of *maʿamoul* and cookies, or hoping to "innovate the family business", as with Refaat's daughter, who has pushed her father to make more vegan products.

These novel models have seen concurrent shifts towards pre-sealed packages. The life coach bemoaned the adoption of "foreign packaging and modernisation techniques" employed by the Hallab companies, which he said have overtaken local techniques and caused the loss of knowledge built up over hundreds of years.[11] This is a familiar tale of the abandonment of indigenous preservation methods. The plastic-sealed packages serve to further commodify the sweet, inviting a different kind of exchange (an Abdul Rahman Hallab senior employee called it "grab and go") and latent disenchantment—as the cultural critic Thomas Hine has argued, "packages promise predictable, risk-free satisfaction, without all the unpredictability and irritation that accompany dealing with other people. And once people are isolated in this way, the society will become less resilient" (1995: 267). A manager at Abdul Rahman Hallab expressed his frustration to me that the new Hallab brand chocolate was not receiving the recognition he believed it was due. He was convinced it was not a matter of flavour but of packaging and labelling, and was working to craft something more compelling.

6.5 Conclusion

This study set out to answer why Tripoli is renowned for sweets and why this renown has endured in the face of urban change. The city's production and trade of sweet goods goes back centuries, prevailing even as key ingredients ceased to be cultivated locally through successful commodification for the modern period. Sweets can be seen as a system of communication; in social life they occupy an important place in hospitality and gift exchange, as well as in signifying status and conferring in-group identity. Similarly, in the Muslim and Christian calendars they are characteristic features of holidays, with hugely symbolic resonance. I suggest their popularity in these events is inscribed through repeated usage, whereby they take on anticipatory value for the next occasion. Their commodification has been furthered through the use of shrewd marketing and innovation of items for changing tastes. Sweets provide a contrast to the images of violence and radicalisation that the city frequently conjures in Beirut and further afield, but struggle to counteract them for locals and outsiders. As a focus of study they offer a way of reckoning with the city's continuities and ruptures, and prove to be revealing of the nature of modern urban life and its distinct needs.

[11] OT, interview.

References

Armstrong, Martin. 2012. Hallab Satisfies Lebanon's Sweet Cravings. *The Daily Star*, August 18.

Barthes, Roland. 1961. Pour une psycho-sociologie de l'alimentation contemporaine. *Annales. Économies, sociétés, civilisations* 5: 977–986. English Edition. 2008. Toward a Psychosociology of Contemporary Food Consumption. In *Food and Culture: A Reader*, ed. Carole Counihan and Penny Van Esterik, 28–36. New York: Routledge.

Cochrane, Paul. 2019. Can You Spare $11.6 Billion? Lebanon's Loans and Luxury Car Sales Paradox. *Counterpunch*, March 29. https://www.counterpunch.org/2019/03/29/can-you-spare-11-6-billion-lebanons-loans-and-luxury-car-sales-paradox/. Accessed 1 May 2019.

Davidson, Alan. 2014. *The Oxford Companion to Food*. Ed. Tom Jaine. Oxford: Oxford University Press.

Deeb, Lara, and Mona Harb. 2013. *Leisurely Islam: Negotiating Geography and Morality in Shi'ite South Beirut*. Princeton: Princeton University Press.

Ferguson, Priscilla Parkhurst. 2014. The French Invention of Modern Cuisine. In *Food in Time and Place: The American Historical Association Companion to Food History*, ed. Paul Freedman, 233–252. Berkeley, Calif.: University of California Press.

Gulick, John. 1967. *Tripoli, a Modern Arab City*. Cambridge, Mass: Harvard University Press.

Heine, Peter. 2018. *The Culinary Crescent: A History of Middle Eastern Cuisine*. London: Gingko Library.

Heinz, Bettina, and Ronald Lee. 1998. Getting Down to the Meat: The Symbolic Construction of Meat Consumption. *Communication Studies* 49: 86– 99. https://doi.org/10.1080/10510979809368520

Helou, Anissa. 1994. *Lebanese Cuisine*. London: Grub Street.

Helou, Anissa, and Linda Pugliese. 2015. *Sweet Middle East: Classic Recipes, from Baklava to Fig Ice Cream*. San Francisco: Chronicle Books.

Hine, Thomas. 1995. *The Total Package: The Evolution and Secret Meanings of Boxes, Bottles, Cans, and Tubes*. Boston: Little, Brown.

Holtzman, Jon D. 2006. Food and Memory. *Annual Review of Anthropology* 35: 361–378. https://doi.org/10.1146/annurev.anthro.35.081705.123220.

Hunt, Lynn. 2008. *Measuring Time, Making History*. Budapest: Central European University Press.

Klein, Jakob A. 2014. Introduction: Cooking, Cuisine and Class and the Anthropology of Food. In *Food Consumption in Global Perspective: Essays in the Anthropology of Food in Honour of Jack Goody*, ed. Jakob A. Klein and Anne Murcott, 1–24. Basingstoke: Palgrave Macmillan.

Larkin, Craig, and Olivia Midha. 2016. The Alawis of Tripoli. In *The Alawis of Syria: War, Faith and Politics in the Levant*, ed. Michael Kerr and Craig Larkin, 181–204. New York, NY: Oxford University Press.

Leeds-Hurwitz, Wendy. 2016. Epilogue: Urban Foodways as Communication and as Intangible Cultural Heritage. In *Urban Foodways and Communication: Ethnographic Studies in Intangible Cultural Food Heritages around the World*, ed. Casey Man Kong. Lum, Marc de Ferrière, and Le. Vayer, 185–200. Lanham: Rowman & Littlefield.

Lum, Casey Man Kong, and Marc de Ferrière le Vayer. 2016. At the Intersection of Urban Foodways, Communication, and Intangible Cultural Heritage. In *Urban Foodways and Communication: Ethnographic Studies in Intangible Cultural Heritages around the World*, edited by Casey Man Kong Lum and Marc de Ferrière le Vayer, 1–22. Lanham: Rowman & Littlefield.

Mauss, Marcel. 1925. Essai sur le don. *L'Année Sociologique*, English Edition. 1990. *The Gift*. New York: W.W. Norton.

Mintz, Sidney W. 1986. *Sweetness and Power: The Place of Sugar in Modern History*. New York: Penguin Books.

Mintz, Sidney W., and Christine M. Du Bois. 2002. The Anthropology of Food and Eating. *Annual Review of Anthropology* 31: 99–119. https://doi.org/10.1146/annurev.anthro.32.032702.131011.

Mnajjed, Safouh. 2013. Al-Tal: Saḥa Lam Takun Tanām. *Attamaddon Newspaper*. http://sawtalbilad.com/old/article.php?idAr=150. Accessed 01 May 2019.

Ohnuki-Tierney, Emiko. 1999. We Eat Each Other's Food to Nourish Our Body: The Global and the Local as Mutually Constituent Forces. In *Food in Global History*, ed. Raymond Grew, 240–271. Boulder, Colo.: Westview Press.

Salamandra, Christa. 2004. *A New Old Damascus: Authenticity and Distinction in Urban Syria*. Bloomington: Indiana University Press.

Terrio, Susan J. 2000. *Crafting the Culture and History of French Chocolate*. Berkeley; London: University of California Press.

Trubek, Amy B. 2008. *The Taste of Place: A Cultural Journey into Terroir*. Berkeley, Calif.: University of California Press.

UN ESCWA. 2015. Poverty in Tripoli. United Nations Economic and Social Commission for Western Asia. https://www.unescwa.org/news/poverty-tripoli. Accessed 1 May 2019.

Zubaida, Sami. 2014. Drink, Meals and Social Boundaries. In *Food Consumption in Global Perspective: Essays in the Anthropology of Food in Honour of Jack Goody*, ed. Jakob A. Klein and Anne Murcott, 209–223. Basingstoke: Palgrave Macmillan.

Henry Peck is a researcher with interests in foodways, urbanism, and human rights. He holds a master's degree in Modern Middle Eastern Studies from the University of Oxford and a BA in International Relations from Brown University.

Part II
Law, Power, and Media

Chapter 7
"An Act Authorizing Sterilization of Persons Convicted of Murder, Rape, Chicken Stealing…": Southern Chicken Theft Laws as an Expression of Racialised Political Violence

Daniel Thoennessen

Abstract This chapter examines the history of laws surrounding chicken theft in the American South, arguing that harsh state punishment of this act served as a form of racist structural violence targeted at the region's African American population. The paper frames this discussion around a 1929 Missouri General Assembly bill that called for the "sterilization of persons convicted of murder, rape, chicken stealing, automobile theft, highway robbery, bombing, mental defectives, epeleptics [sic], and persons afflicted with venereal diseases". Although singling out chicken stealing for such extreme punishment alongside several violent crimes may seem strange to modern eyes, its presence is no aberration. A survey of chicken theft criminalization across three phases of American history reveals that the laws (or lack thereof) punishing this act frequently utilized and reinforced racist stereotypes associating African Americans with chicken theft as a way to target and control this group. State-sanctioned punishments included: slavery's unrestricted physical violence; Reconstruction's disenfranchisement of convicted poultry thieves; and ultimately Progressive Era attempts at sterilization. This chapter proposes that, while specific chicken theft punishments shifted in accordance with their eras' politically acceptable forms of racialised violence, they are unified by their discriminatory intent and application.

Keywords Southern history · Food history · Reconstruction · Eugenics · Chicken theft

7.1 Missouri House Bill No. 290

To modern eyes, the events of January 29, 1929 may seem like a historical aberration. On this date, Missouri State Representative George F. Ballew introduced a bill in the Missouri General Assembly (1929a) calling for the "sterilization of persons

D. Thoennessen (✉)
Department of Nutrition and Food Studies, New York University, New York, USA
e-mail: drt249@nyu.edu

S. Stano and A. Bentley (eds.), *Food for Thought*, Numanities - Arts and Humanities in Progress 19, https://doi.org/10.1007/978-3-030-81115-0_7

convicted of murder, rape, chicken stealing, automobile theft, highway robbery, bombing, mental defectives, epeleptics [sic], and persons afflicted with venereal diseases". One crime stands out immediately. While current sensibilities would decry the use of sterilization as punishment for *any* crime (much less medical conditions and disabilities), it seems particularly strange to see chicken theft singled out for this harsh punishment alongside violent crimes such as murder, rape, or bombing. Legislation targeting this particular act, however, was far from a radical outlier in American history. Rather, Rep. Ballew's bill—which was ultimately voted down by a relatively narrow margin (Lael et al. 2007; Missouri General Assembly 1929b: 1512)—was less of an aberration than a culmination. Tracing the laws surrounding chicken theft throughout the history of the American South reveals that the criminalization and harsh punishment of this act served as a form of racist structural violence targeted at the region's African American population. While specific chicken theft punishments shifted in accordance with their eras' politically acceptable forms of state-sanctioned violence, they are unified by their discriminatory intent and application.

7.2 A Foundation of Racist Stereotypes

The disproportionate criminalization of chicken theft rests upon a long history of racist stereotypes associating this act with African Americans. From the bird's arrival in North America alongside the first European colonizers until well into the twentieth century, the low cost, relative ease, and productivity of raising chickens made them a valuable resource for economically marginalized groups (Loewald et al. 1958). This held true for the enslaved and free Africans living under the brutal exploitation and abuse of American slavery. Subjected to a plantation system in which most were forced to subsist on meagre rations of salt pork and cornmeal that were "objectively deficient in providing…enough sustenance to reproduce their labor", enslaved Africans faced frequent hunger (Opie 2010; Lichtenstein 1988: 417). Access to chickens represented survival. For those with permission, this access came via personal gardens where they raised the low-maintenance birds, providing enslaved people with eggs, meat, and even circumscribed access to an informal economy through the trade or sale of surplus chickens (Williams-Forson 2006). For those denied this opportunity, survival occasionally meant stealing[1] the plentiful and portable birds. Though the combination of "the drive of hunger and a sharp awareness of the plenty around them" made the appropriation of *all* kinds of food both common and justifiable on Southern plantations, enslaved Africans "recalled chickens…as frequently stolen items" (Lichtenstein 1988: 417). "I used to steal some chickens 'cause we didn't have enough to eat…and I didn't think I done wrong,

[1] Though I will use the terms "stealing" and "theft" to match the language of existing discussion and scholarship around these acts, these terms hardly carry the proper moral or ethical connotations to describe enslaved Africans' justifiable appropriation of "masters'" property under an abusive and exploitative system.

'cause the place was full of 'em", remembered one formerly enslaved person in George Rawick's collection of slave narratives (as cited in Lichtenstein 1988: 417). Famed educator Booker T. Washington (2004) similarly recalled the memory of his mother feeding her hungry children a stolen chicken, saying "no one could ever make me believe that my mother was guilty of thieving. She was simply a victim of the system of slavery" (3).

Though African Americans were far from the only group stealing chickens,[2] the same racist system that made this theft necessary ensured that they became uniquely associated with it. Even if, as Williams-Forson (2006) suggests is probable, "theft by blacks was less common than believed" (26), frequent reports of these crimes reflected and reinforced antebellum whites' belief that enslaved people had "inherent racial thieving tendencies" (Henry 1914: 81). By the 1800s, chicken had been singled out as a particular target of this supposed disposition toward theft, and "black men and women had fully come to be associated with chicken stealing" (Williams-Forson 2006: 27). The eventual prevalence and strength of this racist association is made clear by a *New York Times* article[3] entitled "The Chicken Question" (1882). The article uses the arrest of "a colored man...in possession of felonious chickens" to question why "a total inability to resist temptation in connection with Caucasian hen-roosts characterizes the negro [sic]...there is evidently an affinity between colored men and chickens so strong as to defy the efforts of morality to keep the two separate" (para. 1–4). By this point in American history, discussing chicken theft carried a host of implications about exactly *who* was stealing the chickens. It seems likely that those arguing for particularly harsh punishment of this crime had a similar target in mind.

In a Southern political structure designed to maintain white social, political, and economic dominance, the strength of these racist associations made chicken an ideal locus of power and control. The prejudicial intent and application of chicken theft laws gave the state the opportunity to "control and the define the right to sustenance" (Lichtenstein 1988: 417), limiting access to a necessary food source for African Americans while simultaneously creating a legal mechanism to legitimate their disproportionate and often arbitrary punishment. As is often the case in a system built on this type of structural violence, these damaging effects spread far beyond those the laws were ostensibly written to target. Conclusive evidence of guilt was not necessarily a relevant factor in the prosecution of chicken theft. Purported thieves "were often convicted even when the evidence against them was preposterous" (Williams-Forson 2006: 27)—as in the 1876 case of an African American woman in Virginia sentenced to thirty-nine lashes based on the "testimony" of a mother chicken who somehow convinced a court that the stolen birds in question were her offspring (Rude 2016). Such arbitrary prosecution extended the fear of punishment well beyond those guilty of theft, further demonstrating the way unequal application of laws surrounding

[2] Chickens were eaten across nearly all racial and class lines—Williams-Forson (2006) even notes that enslaved Africans were often the ones raising and cooking chicken specifically for the consumption of white slave owners.

[3] Likely either an op-ed or a letter to the editor.

this act weaponised racial stereotypes to attack their desired targets. Though identifying and accounting for all the interactions of race, class, and gender that occur when discussing such a widely consumed food inevitably presents complications,[4] it ultimately seems that the pernicious exploitation of specific racial stereotypes made the harsh criminalization of chicken theft a method of targeted structural violence against African Americans. Tracing state punishment of this crime throughout American history reveals that, while the accepted expression and reasons for this violence may have shifted over time, its racist foundations remained remarkably stable.

7.3 Chicken Theft Punishment Under Slavery

An early glimpse of these racist foundations can be seen in the general absence of specific legal statutes regarding chicken theft during the era of slavery in the American South. Despite firmly established stereotypes linking both free and enslaved Africans to chicken theft, and concern among slave owners that "chickens were disappearing at the hands of black people" (Williams-Forson 2006: 27), there appears to be a notable lack of laws singling out chicken theft for particular punishment in the years prior to Emancipation. It seems that, though the ruling white planter class sought to police, punish, and subjugate the enslaved population in whatever ways maximized the exploitation of their labour while minimizing the persistent threat of slave rebellion (Dormon 1977), the direct and all-encompassing racism of state-sponsored slavery rendered such specific, indirect legal targeting of African Americans unnecessary. Violent racism against African Americans was a fundamental and legal part of nearly all social, political, and economic structures in the antebellum South. Enslaved Africans were considered property with no human rights in the eyes of the law, providing "masters...the legal invitation to dehumanize their slaves" (Fede 1992: ix-x). In practice, this lack of legal boundaries meant that slave owners seeking to control this population's access to food—or use accusations of its theft as a means of punishment—could do so at will via state-legitimized physical violence. As Lichtenstein (1988) notes, slave owners' responses to theft reflected this, with the accused[5] often facing "brutal punishments...many descriptions of theft found in the slave narratives and testimony are followed by references to the whippings received by those slaves unlucky enough to be caught" (417).[6] In a political environment that condoned this level of open and pervasive racist abuse and dehumanization, there was seemingly little need for laws that indirectly pursued similar goals—particularly if they targeted a group with no legal rights in the first place. The absence of specific chicken theft laws during this period—and their later emergence—suggests that they

[4] Particularly because chicken also represented an important food source for poor white Americans.

[5] Who had no legal recourse for defending themselves against spurious accusations.

[6] The reach of this oppressive institution ensured that even free African Americans and fugitives from slavery were subject to such threats, as both faced the fear of apprehension that could possibly subject (or re-subject) them to imprisonment or bondage based on minimal evidence.

were this type of indirect attack, exploiting stereotypes to place a thin veneer of legitimacy (and plausible deniability) between the laws and their more explicitly racist intentions.

7.4 Chicken Theft Laws Under Reconstruction and the Early Jim Crow South

The Civil War and the emancipation of enslaved Africans marked a drastic shift in the South's politics and power structure, ushering in a new political climate that allowed the legal weaponisation of chicken theft stereotypes to take hold and accelerate. As Williams-Forson (2006) notes, the Federal Government's military victory and subsequent dismantling of legal slavery did not diminish racism among Southern whites. Instead it fuelled it, as the large population of newly free African American men with citizenship and voting rights represented a significant threat to whites' formerly firm grasp on political power. Furthermore, the recently imposed Fourteenth Amendment prevented the kind of overt denial of legal rights permissible under slavery. An exception existed, however, permitting disenfranchisement "for participation in rebellion, or other crime" (U.S. Const. amend. 1868, XIV: § 2). This meant Southern Democrats[7] who wished to deter or reduce the number of African American voters—who "voted disproportionately for the Republican party"—simply had to tailor their states' criminal disenfranchisement laws to apply this punishment to what Holloway (2009) identifies as "the crimes that African Americans were thought likely to commit" (934–935). The stereotypes of African Americans as predisposed to livestock and poultry theft made these crimes an ideal vehicle for this discrimination. As a result, by the 1880s—once federal enthusiasm for intervening to protect African Americans' civil rights had notably diminished—nearly every Southern state had "expanded disenfranchising offenses to include petty crimes", with a specific and often explicit focus on forms of livestock theft (Ibid.: 934, 948). *Alabama legislative acts* (1875) highlights several examples of such statutory changes. The state's legislators elevated theft of a "cow...hog, sheep, goat, or any part of any outstanding crop of corn or cotton" to the category of felony, making them punishable by jail time and a resulting loss of voting rights (Ibid.: 260). Similarly, the definition of felonious burglary was expanded to include breaking into "any building, structure or inclosure [sic]" near a dwelling (Ibid.: 258)—such as the enclosed "Fowl Houses" that Stoddard's (1879) contemporaneous guidebook *Poultry Architecture* recommended for raising chickens.

Though Rep. Ballew's singling out of chicken theft was still well in the future, statements of the time reveal that targeting chicken theft was central to the racist intent and application of these Reconstruction and early Jim Crow Era livestock laws. Both sides of the debate surrounding such legislation acknowledged that a

[7] Who represented the dominant white planter class that was eager to reassert its control.

primary goal of criminalizing petty livestock theft was the imprisonment and disenfranchisement of African Americans. Southern Democrats often couched their racist political goals within the language of high-minded personal and civic responsibility, arguing that African Americans' supposed inability to resist chicken demonstrated they "were creatures of desire and emotions, not men of morals and restraint" and could not be trusted with the responsibilities of suffrage (Holloway 2009: 958). "What sort of claim to participation in the matter of governing the country has a chicken thief?", asked an 1876 Virginia newspaper article ("Exactly Right" 1876: 2). Others, however, were even more up front about lawmakers' ultimate motivations and the importance of punishing chicken theft to achieve these goals. An article in Mississippi's *Greenville Times* (1876) following the passage of the state's "Pig Law" (which punished livestock theft with disenfranchisement but notably excluded direct mention of chicken) joked that the new "legislation would deprive Republicans of 'fresh meat'" in the voting booth, but that African Americans should "Bless the Lord, they haven't said a word about Chickens!" (1). These intentions were obvious to opponents of these bills such as African American attorney Charles N. Otey, who testified before the Senate in 1880 that "I believe myself, as a Republican, that the law was made for the purpose of disenfranchising colored men" (as cited in Holloway 2009: 947). The application and consequences of these laws supported this belief. Though the livestock laws were ostensibly colour-blind, and Americans of all racial and ethnic categories stole chickens, it seems that African Americans were disproportionately arrested and punished for this crime—often with evidence as ludicrous as the case of the woman convicted based on a chicken's testimony (Williams-Forson 2006). African Americans with petty theft convictions were regularly turned away at polling places all over the South[8]—with notably increased enforcement during close elections—yet there is "little evidence of selective enforcement" against lower-class whites with similar convictions who *should* have been subject to the same punishment (Holloway 2009: 948). While chicken theft was not yet the sole legal focus of such discriminatory laws, the political utility of exploiting racist chicken theft stereotypes to target Africans Americans was abundantly clear.

7.5 Chicken Theft Laws During the Progressive Era

Rep. Ballew's bill represents a Progressive Era culmination of this increasingly specific focus on chicken theft as a vehicle for targeted political violence against African Americans. Much like the transition from slavery to Reconstruction, the rise of the Progressive Era in the early twentieth century[9] brought with it a new political climate and accompanying expressions of racist structural violence. While the

[8] As was often the case when it came to the application of racist Jim Crow laws, it is also likely that many African Americans were turned away from the polls for having the same or a similar name as someone with this type of conviction on their record.

[9] Which overlapped with the Jim Crow Era in the South.

American South of this time period was characterized by continued legal segregation and the general denial of African Americans' civil rights, this era saw the peak of a specific form of intellectual justification for this oppression—scientific racism (Aldrich 1979). A blend of Social Darwinist ideas about white racial superiority and growing faith in the seeming objectivity of natural sciences (Dennis 1995), scientific racism was "racial science premised on racial hierarchy—the idea that different races can be ranked by innate fitness" (Leonard 2003: 689). The popularity of these ideas was expressed politically through the growing eugenics movement, which sought to promote reproduction of those it considered genetically superior while discouraging the reproduction and even promoting the sterilization of populations it deemed unfit—including numerous racial groups, criminals, and the "feebleminded" (Rosen 2004: 166). Once again, chicken theft laws and their discriminatory application served as a useful way for those in power to target particular demographics for this harsh punishment. The beginning of the twentieth century saw the American eugenics movement aggressively push for policies around the country that sterilized a wide range of criminals—often including "'genetically inferior' chicken thieves" (Rude 2016: loc. 2388). Their success is demonstrated by state sterilization programs like Oklahoma's Habitual Criminal Sterilization Act of 1935, under which chicken theft counted as one of the "felonies involving moral turpitude" (a category which notably did not include white-collar crimes like embezzlement) that justified sterilization for repeat offenders (*Skinner v. Oklahoma* 1942).[10] This harsh punishment reflected some of the new fears of the dominant white political class. While convicted thieves were still prevented from voting, the effect of Jim Crow laws and violent intimidation across the South meant that "African American disfranchisement was so extensive and Democratic rule so complete that disfranchising these individuals contributed little to the larger racial and partisan agenda" (Holloway 2009: 936). Adding the existential violence of state-sanctioned sterilization on top of disenfranchisement hints at the white ruling class's growing fear of losing their political, economic, and cultural dominance due to demographic change, miscegenation, and so-called "race suicide" (Aldrich 1979: 1).

While its narrow focus on chicken theft makes it an even more specific attack on this crime than many of its contemporaries, Rep. Ballew's 1929 call for the sterilization of persons convicted of chicken stealing is ultimately more of a distillation of the era's predominant attitudes and punishments than a departure from the norm. It is difficult to say with certainty that Rep. Ballew included chicken theft in his bill to specifically target African Americans for sterilization alongside the "mental defectives, epeleptics [sic], and persons afflicted with venereal diseases" that were common targets of the eugenics movement (Missouri General Assembly 1929a). In practice, however, generations of racist stereotypes and prejudicial application of earlier laws would likely ensure this community would be disproportionately apprehended and punished regardless of his intent. While poor white chicken thieves would likely be

[10] *Skinner v. Oklahoma* would later overturn this case because of this unequal application of punishment for similar crimes.

included in those the eugenics movement would consider unfit for reproduction,[11] images from the era's burgeoning popular media make it clear that one type of person continued to fit the public profile of a chicken thief. The trope of African Americans as "lazy shiftless chicken-stealers" appears again and again in advertisements, postcards, and other cultural products[12] of the time (Williams-Forson 2006: 45). One Missouri whiskey advertisement from the years prior to Ballew's bill depicts an African American character with a stolen chicken in one arm and a watermelon in the other, struggling to decide which object to drop in order to pick up a free bottle of whiskey on the ground. The caption reads "I'se in a perdickermunt", trusting its audience to understand the racist joke (Mitenbuler 2015: 147). The prevalence of such images suggests that, by specifically calling for the sterilization of chicken thieves, Rep. Ballew and the bill's other supporters likely had a similar picture in mind as the target of this proposed punishment.

7.6 Conclusion

After looking back across the history of chicken theft stereotypes, laws, and punishments in the American South, Rep. Ballew's classification of this crime alongside murder, rape, and bombing no longer seems quite so out of place. While the crime itself is typically far from violent, the long-held association of chicken theft with African Americans meant the presumed criminal represented a significant threat to those seeking to preserve a racist political and economic system. As a result, the punishment of this crime served as a state-sanctioned response to this perceived threat. Tracing these punishments across Southern history reveals that, while Ballew's bill is notable for its specificity, its harsh treatment of chicken thieves was not unique. Whether it was the physical violence of slavery, Reconstruction Era disenfranchisement, or Progressive Era attempts at sterilization, the discriminatory intent and unjust application of Southern chicken theft laws and punishments has long served as a tool of racist structural violence against the region's African American population.

[11] The convicted chicken thief who was castrated under the Oklahoma law and challenged its constitutionality in *Skinner v. Oklahoma* appears to have been Caucasian—though it would not be unusual for those seeking to challenge a case to select a somewhat non-representative plaintiff with the goal of maximizing their chances in court.

[12] Including the massively popular film Birth of a Nation (1915), which included a glamorized depiction of the history of the Ku Klux Klan during reconstruction and used fried chicken as a racist symbol of African Americans' supposed gluttony and unworthiness for political participation.

References

Alabama legislative acts, 1874–1875 [part 2 of 3]. 1875. Montgomery: Alabama Department of Archives and History. http://digital.archives.alabama.gov/cdm/compoundobject/collection/legislature/id/25986/rec/77. Accessed 30 April 2020.

Aldrich, Mark. 1979. Progressive economists and scientific racism: Walter Willcox and black Americans, 1895–1910. *Phylon (1960-)* 40: 1–14.

Birth of a Nation. 1915. Directed by D.W. Griffith. USA: David W. Griffith Corp. https://www.youtube.com/watch?v=GBzDH-Vwzy4&t=37s. Accessed 15 May 2019.

Comments on this Act. 1876. *Greenville Times*, April 15.

Dennis, Rutledge M. 1995. Social Darwinism, scientific racism, and the metaphysics of race. *The Journal of Negro Education* 64 (3): 243–252.

Dormon, James H. 1977. The Persistent Specter: Slave Rebellion in Territorial Louisiana. *Louisiana History* 18(4): 389–404. https://www.jstor.org/stable/4231728. Accessed 30 April 2020.

Exactly Right. 1876. *Richmond daily dispatch*, November 4.

Fede, Andrew. 1992. *People without rights: An interpretation of the fundamentals of the law of Slavery in the U.S. South*. New York City: Routledge. https://doi.org/10.4324/9780203815588.

Henry, Howell Meadoes. 1914. *The police control of the Slave in South Carolina*. Emory, VA. https://archive.org/details/policecontrolofs00henr/page/n10. Accessed 30 April 2020.

Holloway, Pippa. 2009. A chicken-stealer shall lose his vote: Disfranchisement for Larceny in the South, 1874–1890. *The Journal of Southern History* 75(4): 931–962. http://www.jstor.org/stable/27779119. Accessed 30 April 2020.

Lael, Richard L., Barbara Brazos, and Margot Ford McMillan. 2007. *Evolution of a Missouri Asylum: Fulton State Hospital, 1851–2006*. Columbia: University of Missouri Press.

Leonard, Thomas C. 2003. 'More merciful and not less effective': Eugenics and American economics in the progressive Era. *History of Political Economy* 35(4): 687–712. https://muse.jhu.edu/article/50841. Accessed 30 April 2020.

Lichtenstein, Alex. 1988. "That disposition to theft, with which they have been branded": moral economy, slave management, and the law. *Journal of Social History* 21(3): 413–440. https://www.jstor.org/stable/3787592.

Loewald, Klaus G., Beverly Starika, Paul S. Taylor, and Johann Martin Bolzius. 1958. Johann Martin Bolzius answers a questionnaire on Carolina and Georgia: Part II. *The William and Mary Quarterly* 15 (April): 228–252.

Missouri General Assembly. 1929a. *An Act Authorizing sterilization of persons convicted of murder, rape, chicken stealing, automobile theft, highway robbery, bombing, mental defectives, epeleptics, and persons afflicted with venereal diseases*. HB No. 290. 54th House Session. Introduced in House January 29, 1929. In author's possession.

Missouri General Assembly. 1929b. *Journal of the House of the State of Missouri, Volume II*. 54th General Assembly, Ninetieth Day, April 24, 1929. In author's possession.

Mitenbuler, Reid. 2015. *Bourbon empire: The past and future of America's Whiskey*. New York: Penguin Books.

Opie, Frederick Douglass. 2010. *Hog and hominy: Soul food from Africa to America*. New York City: Columbia University Press. Kindle.

Rosen, Christine. 2004. *Preaching eugenics: Religious leaders and the American eugenics movement*. New York City: Oxford University Press.

Rude, Emelyn. 2016. *Tastes like chicken: A history of America's favorite bird*. New York: Pegasus Books. Kindle.

Skinner v. Oklahoma Ex Rel. Williamson, 316 U.S. 535 (1942).

Stoddard, H.H. 1879. *Poultry architecture: How to build handsome and convenient fowl houses durably and economically*. Hartford, Connecticut. https://books.google.com/books?id=Mso9AAAAYAAJ&pg=PA28&source=gbs_selecte_pages&cad=3#v=onepage&q&f=false. Accessed 29 April 2020.

The Chicken Question. 1882. *The New York tmes*, April 15. https://www.nytimes.com/1882/04/15/archives/the-chicken-question.html. Accessed 12 May 2019.

U.S. Const. amend. XVI., § 2. 1868.

Washington, Booker T. 2004. *Up from Slavery: An autobiography*. West Berlin, NJ: Townsend Press. Kindle.

Williams-Forson, Psyche A. 2006. *Building houses out of chicken legs: Black women, food, and power*. Chapel Hill: The University of North Carolina Press. Kindle.

Daniel Thoennessen is a 2020 graduate of New York University's Food Studies Master's program with a particular interest in the foodways and history of the American South. In addition to his academic studies, he is currently directing and editing *Holy Bird*, an in-production documentary film examining the cultural and culinary history of fried chicken in the United States. He lives in New York City.

Chapter 8
Free Breakfast and Taco Trucks: Case Studies of Food as Rhetorical Homology in Political Discourse

Suzanne Cope

Abstract This chapter examines examples from the last half century of food as a rhetorical homology—equating food culture and activism and political discourse. The Black Panther Party's Free Breakfast for Children Program is compared to the contemporary indigenous group the I-Collective, whose identity and activism feature food-focused content both virtually and in person. The two case studies illuminate the power of food as a broader rhetorical tool brandished today by activists and politicians alike. While food has long been used as a political tool, the Black Panther Party's Free Breakfast for Children Program is one of the most potent examples of food involving explicit political discourse. Their 1969 national anti-hunger initiative began as a way to make a positive difference in the lives of marginalised children, but also functioned to communicate values, intention, and political ideology, linking discourses and creating what Barry Brummett called a "rhetorical homology". With the rising influence of social media, and the popularity of food on these platforms we see this happening again in modern discourse. The indigenous activist group I-Collective is a contemporary example of a group using food as political rhetoric in a similarly explicit way as did the Black Panther Party a half century earlier.

Keyword Black Panther Party · I-Collective · Food studies · Rhetoric · Politics · Communication · Instagram · Rhetorical homology

8.1 Brief History of Food as Political Rhetoric

In such historic examples as the Boston Tea Party, Ghandi's Salt March, or the suffrage slogan "Bread and Roses!", edible (and potable) ideas and imagery have long been employed to make political statements beyond the realm of the table. While these enduring ideas were just as often claimed by male activists as by female, because the act of feeding others has long been relegated as women's work, until the second half of the twentieth century, the politics of food on the table has almost

S. Cope (✉)
Expository Writing Program, New York University, New York, USA
e-mail: suzanne.cope@nyu.edu

S. Stano and A. Bentley (eds.), *Food for Thought*, Numanities - Arts and Humanities in Progress 19, https://doi.org/10.1007/978-3-030-81115-0_8

exclusively been in women's domain. Early soup kitchens or free school lunches in the United States were primarily founded and staffed by women, often—but not exclusively—inspired by religious convictions. And many of these programs, and increasingly the schools that took over the feeding of students, also had ulterior motives, seeking to assimilate the poor immigrants they were serving according to their own biased perspective of appropriate food consumption. The free food they offered had larger agendas and expectations of culture, assimilation, and political and social power (Carstairs 2017; Ruiz 2015; Shapiro 1986). Yet few examined these free or inexpensive meals through the lens of political messaging.

Further, from the late 1800s through mid-twentieth century, writing about food in mass media was also largely the purview of women. Often relegated to the "women's pages", articles about food were not just recipes and family advice, but conversations on poverty, nutrition, and public policy, and progressive attitudes around immigration and culture. They had relatively little impact beyond their core readership, however, in part because men rarely took notice of these sections. Yet these sections were important as they allowed women to feel politically activated and to demonstrate agency with regard to changing social mores and expectations (Cope 2018; Voss 2014).

As food journalism migrated from women's pages in mid-century, it grew in popularity.[1] By the late 1960s as food writing became more mainstream, the women's pages proved so popular that men were now writing for them and, not surprisingly, co-opting the editing. The shift facilitated the loss not just of women's perspectives, but also much of the politically-inflected food conversations as well. A number of markers demonstrate this growth, including the food media's developing prestige and broadening market, the rise of food television, and the emergence of food-centric lifestyle influencers (Cope 2018; Voss 2014).

Food, however, had long been considered a weapon of power and political change by government and other forces, despite minimal recognition in public debates in the media and public forums. The US government itself had been employing food as a tool in foreign and domestic policy—providing free food to impoverished politically volatile groups to keep them mollified, for example, as was done with the Bolsheviks after World War I—in the hopes of quelling revolution. Conveniently, this politically strategic food relief allowed the United States opened up trade routes using the established avenues for distribution (Radosh 2011). Free food had become politically and economically beneficial—but, for obvious reasons, the government did little to make this part of the broader public discourse. This is but one example of powerful groups either providing food or enforcing its absence to affect political sentiment. Henry Kissinger was often quoted as saying, "Control oil and you control nations; control food and you control the people". Yet in mid-twentieth century US few people gave much thought to the power of food, either as a material thing or a form of rhetoric, to politically educate or change public sentiment.

---

[1] Left unsaid here due to scope and space are the implication of gender and power inherent in this shift.

In the mid 1960s, however, the Black Panther Party revealed and transformed the discourse. Originally called the Black Panther Party for Self Defense, this Oakland, California community activist group is most commonly known for protesting and challenging rampant police brutality. Although their manifesto, called the Ten Point Program, pronounced their mission as fighting for access to healthy food, housing and health care, early media coverage characterized the Panthers as militant and potentially violent extremists. This narrative affected public support not only nationally, but even among those they sought to serve in the mainly inner city neighbourhoods. For example, a 1968 *Time* magazine article titled "The Panther's Bite" described the Panthers as "not only militant but also militaristic". Asserting the Black Panthers as being "A state of war", and "devoted to some hard-line Chinese Communist doubletalk" the article accuses them of taking part in burglaries, coercion, and armed offenses ("The Panthers' Bite" 1968).

However, when in fall 1968 they began their Free Breakfast for Children Program, which quickly expanded from Oakland across the country, the Black Panthers' popularity rose noticeably both among the locals they served as well as other social welfare organizations. The Panthers' people-focused "survival program" initiatives as they called them (of which the Breakfast program was the centrepiece), changed not only community members' perceptions of the Panthers themselves, but the Breakfast program also helped educate these communities politically. This political education occurred in part through their empowering of children, whom they helped to honour their heritage and feel pride for people of African descent in general. The Panthers also demonstrated their politics through the modelling of socialist ideals, especially their perspective that children should not go hungry, especially in the richest country in the world (Heynen 2009; Murch 2010).

It is important to note that the Black Panther Party did not establish the Free Breakfast for Children Program or any other community program as a means to manipulate public perception. Rather these efforts were part of their core mission, and a reaction to the rightward movement of local and national politics. The Free Breakfast for Children Program both filled a community need—the feeding of hungry children—and provided a model for their socialist ideals. Likewise, the nationally-distributed Black Panther Party newspaper, *The Black Panther*, critiqued government policies around food, health care, and education (among other topics), which provided an additional rhetorical perspective to their on-the-ground activism (Murch 2010). Further, the breakfast program itself demonstrated these ideals in action, which reinforce their perspective in *The Black Panther* and as discussed in the community. Chicago Black Panther Party member Deborah Johnson (who would later be known as Akua Njeri) reflected, "We know that people didn't have an understanding of socialism or communism and that they might say they're against that.... But people basically thought that children had a right to be fed and learn on a full stomach" (Ibid.: 175–176).

As communities and other individuals begin to see the Black Panther Party in a more positive light because of the Free Breakfast for Children Program, more liberal mainstream media outlets similarly began to soften their rhetoric toward the group. A 1970 *Esquire* article, "Is It Too Late For You To Be Pals With A Black Panther", complicated the negative media-created narrative by highlighting the Free

Breakfast for Children program among other BP community work. This article poses the questions, "Aren't the Panthers helping to create a better world for **black** kids? What could be more into the good old American populist tradition than that?".

As a result, the Black Panther Party, however unintentionally, created a "rhetorical homology" (Brummett 2008) that aligned the Free Breakfast for Children Program with a larger rhetorical socialist ideals of supporting community health and well-being. Their discourse about and around the Free Breakfast for Children Program demonstrated the program as a model, creating a "linkage between discourses" (Thompson 2012) that can move beyond that immediate time and space. While the Black Panthers' intent was to feed and educate their local communities, which had the potential to activate localized power, the media amplified this message and thus the rhetorical power of their work.

The media vilification of the Black Panther Party was strongly influenced by the propaganda campaign waged by the FBI's COINTELPRO initiative, led by FBI Director J. Edgar Hoover (Heynen 2009). Even the aforementioned *Esquire* article notes the FBI's role in creating and perpetuating their violent image via the press. The rising popularity of the Free Breakfast for Children Program was seen as so powerful in changing public sentiment of the Black Panther Party that an internal memo from FBI Director J. Edgar Hoover noted the following:

> The Breakfast for Children Program (BCP) has been instituted by the BPP in several cities to provide a stable breakfast for ghetto children. ... The program has met with some success and has resulted in considerable favorable publicity for the BPP. ... The resulting publicity tends to portray the BPP in a favorable light and clouds the violent nature of the group and its ultimate aim of insurrection. The BCP promotes at least tacit support for the BPP among naive individuals ... and, what is more distressing, provides the BPP with a ready audience composed of highly impressionable youths. ... Consequently, the BCP represents the best and most influential activity going for the BPP and, as such, is potentially the greatest threat to efforts by authorities ... to neutralize the BPP and destroy what it stands for. (Churchill and Wall 1990)

It is, perhaps, surprising that a program to feed hungry children was seen as being so politically powerful. This is precisely, however, the demonstration of the power of food as rhetoric: the selfless collection of donated food and the willingness to meticulously plan, organize, and staff a daily free breakfast program for thousands of kids daily around the country was evidence of the Black Panther Party's excellent community organizing skills, their ability to respond to the needs of a community, and the benefit of a socialist political and social framework. Further, the program gave them a platform for positive and productive communications with the community, and a media-genic topic for print, television, and radio.

Through the Free Breakfast for Children Program specifically, and the survival programs more broadly, the Black Panthers were their own best argument for a more socialist society, functioning to convince both their communities and larger allies that a socialist approach to caring for people is both possible and just. They communicated this sentiment in person, in their weekly popular *Panther Paper* distributed around the country for years, and in the multitude of interviews and media coverage of their programs. This can be called the first use of food as a tool of political discourse in modern America that reached a mass audience and represented a larger political ideal. In fact, scholars attribute this program with the expansion of United States anti-hunger policy more broadly. As historian Potorti (2014) assessed, the "Panthers' food programs and anti-hunger politics worked to address the persistence of food insecurity by dramatizing its political roots and implications" (48).

While the Black Panther Party had a goal of feeding hungry kids breakfast, their larger purpose included greater wealth equity, among other revolutionary goals, of which they were only partly successful. Their highly popular and successful survival programs became a template for government-run programs, which helped communities but also reclaimed the power of rhetoric as these programs demonstrated for the Panthers. However, the FBI's COINTELPRO initiative, which wasn't public knowledge until the 1970s, succeed in their efforts to "neutralise" the Black Panther Party through propaganda, spies and rumours, such as providing local police false information to wrongly raid, arrest, and sometimes harm or kill members. Ultimately the costs—both financial and human—of these efforts created too much pressure, leading to internal strife in Panther chapters. The Black Panther Party's influence ultimately dwindled until it became a shadow of its original form (Murch 2010; Potorti 2014).

8.2 Food and Political Rhetoric: 1970–2016

In the ensuing years there were no initiatives that came close to capturing the rhetorical and socio-political power of the Free Breakfast for Children Program, for a number of social, economic, and political factors. First, it is widely agreed that the success of the Free Breakfast for Children Program spurred the government to action to create a federal free breakfast program under the Child Nutrition Act of 1975. Other programs emerged from the growing consciousness of the prevalence of hunger at this time as well, including The Older Americans Act of 1972, which provided funds to feed the elderly and Special Supplemental Nutrition Program for Women, Infants, and Children (WIC), that began as a pilot in 1972 and was given permanent status in 1975. Federal food assistance expenditures soared from \$687 million in 1967 to \$9 billion in 1979 (USDA 2018). Thus the government began addressing the problems around food access that the Black Panther Party helped to highlight, quelling further outcry.

Further, the United States in the 1970s and 1980s witnessed an emergence of the "back to the land" movement (in part as a reaction to the growth of industrial farming) which focused food activism more inwardly on the individual and smaller communal work as opposed to broader political action. While there was some focus on food within the larger environmental movement, through the end of the twentieth century food was not specifically utilised as the catalyst for a non-food-related political conversation (Belasco 2006).

Italy's Slow Food movement (which one might argue did create a rhetorical homology, but not necessarily an explicitly political one) didn't reach the United

States until 2000, and, with the rise of farmer's markets, the artisanal movement in the late aughts, and locavorism, food in this period was shifting toward a more political orientation regarding specific food choices and their larger socio-political, economic, and personal connections. However, this movement was criticised as being expensive and inaccessible for many, and at times performative (Cope 2014; Wexler et al. 2017). Still these movements helped facilitate the modern expression of food as a lens through which one can express identity, and engage in political discussion, among other multi-dimensional discourses, which were amplified by the growth and influence of social media. While there were certainly many socio-political food-related protests, initiatives, and debates happening during this time, arguably none captured both the media and political attention in a way that married community action and larger political discourse, achieving the power of the political rhetorical homology of the Free Breakfast for Children Program.

8.3 The Rise of Social Media: 2016—Present

The rise of digital social media in the aughts and early 2010s fostered the growing influence of food bloggers and social media influencers, with food playing a starring role across lifestyle portals and branding. While social media statistics can be variable, the "#food" moniker has long been a popular aggregator. In fact, in spring 2020 it was the twenty-seventh most popular Instagram hashtag of all time, with almost 400 million tags ("Top 100 Hashtags on Instagram" 2020). Additionally, numerous metrics indicate the popularity of food on social media, the rise of hashtag activism, and the connection between the two. For example, by one metric in October 2019 the top two Instagram influencers overall—Jamie Oliver and David Chang—are both food folks, both of whom espouse a specific political viewpoint at times in their posts (Forsey 2019). Further, the popularity of food as a social media topic and hashtag functions to cultivate armchair or hashtag activism as well as socio-political discussion.

As food's cultural influence grew, so did its use in political messaging. To backtrack chronologically for a moment, with the rise of television as a popular medium in the 1960s, for example, Presidential elections moved away from party identification and toward supporting individual candidates. This so-called "cult of personality" emphasis meant that candidates needed multiple ways to connect with potential voters (Perlman 2013). In the 1990s, due in part to technological advances in voter polling, candidates sought to explicitly connect with middle class Americans, increasingly through food messaging (Ibid.: 44.50). Whether through diner photo ops or the decision to eat fried chicken with their hands versus utensils, candidates have continued to connect with—some say pander to—various constituents through food. This is sometimes to the candidate's detriment as given the rise of social media, even capturing a candidate eating lunch privately can inspire a tweet or post and accompanying commentary.

This trend found its apex (or nadir) during the 2016 presidential race, propelling food messaging beyond news sound bites and critical op eds. Infamously on September 1, 2016, in an MSNBC interview "Latinos for Trump" co-founder Marco Gutierrez quipped, "If you don't do something about it, you're gonna have taco trucks on every corner". This quote was tweeted by the show, @allinwithchris, and a twitter storm ensued. Immediate analysis noted that this comment was a "dogwhistle" to Trump's core supporters, equating taco trucks with negative immigration stereotypes (Lopez 2016). A seemingly off-handed, if racist, comment turned into the rhetorical homology of the taco truck. The backlash was immediate, resulting in the preference for taco trucks becoming shorthand for people's broader political worldview. A search of the hashtag #tacotruck 2 years after the original tweet shows a mix of tweets about trucks that simply serve tacos and political commentary. In Spring 2020 the live algorithm Ritetag.com noted that #tacotruckoneverycorner was used an average of four times per hour, the same as #americangirl—but, by their calculation, earning *48 times* more exposure per hour (Popular hashtags for tacotruckoneverycorner on Twitter and Instagram 2020).

Further evidence indicates that food as a rhetorical tool has become even more prevalent since the 2016 election cycle. After the election there emerged an increase in food-related political action, including fund-raising dinners cooked by refugees or immigrants, or public support of political protests through food donations, the result is an even more intense popular alignment of political and cultural food discourses. In early 2017 even the venerable food writer Pollan (2017) called for likeminded folks "to apply the energy of the food movement to preserving our democracy". One of the most effective current examples of this intersection, in a way reminiscent of the Black Panther Party, is the I-Collective.

8.4 I-Collective

The I-Collective—the "I"s stand for "indigenous", "inspired", "innovative", and "independent"—is a group of indigenous activists from around the U.S., Canada, and Mexico focusing on food sovereignty and political activism through food. Currently consisting of eighteen members, to join one must affirm a set of values including anti-racism language and support for Latinx rights, acknowledging that these groups are also indigenous and separated from Native American Indians by artificially constructed borders.

The I-Collective began in 2017 with three activists—Erica Scott, Neftalí Durán, and Liz Charlebois—who first connected mostly via email and social media. As more members joined, including chefs, artists, photographers, and scholars, the founders insisted on keeping the group's membership a "collective", without established top-down leadership and with a conscious effort to amplify the voices of female members.[2] Thus one of the group's stated goals is to "uplift female leadership" in ways aligned with these traditions (Echo-Hawk 2018; Scott 2018).

The I-Collective activism and anti-colonisation efforts happen both at in-person events that usually involve sharing pre-colonial food, and on social media, most commonly Instagram. This group operates at the intersection of native traditions and values and modern media and popular food culture. The I-Collective strives to amplify the narrative of US government-sanctioned oppression of indigenous groups, through their messaging on Instagram and during food events, among other food activism activities.

While the I-Collective is fighting for indigenous rights more broadly, particularly within North and Central America, most of their members are Native Americans and also strive to keep their traditions alive while highlighting the violent history of European-American colonisation of native populations. The I-Collective reframes the long-held narrative of American history. They note that while the atrocities are too numerous to recount, broadly European colonisers stole native land upon arrival in the "New World", as they called it, and as they continued building settlements murdered many Native Americans and forced many others to relocate to reservations, often far from their ancestral land. On these reservations they restricted their foodways, among other expressions of their culture. They stand in solidarity with the Black Lives Matter movement as well, amplifying the history of enslaved people in the Americas and supporting current issues and causes, and working together for food sovereignty for people of colour. Much of the I-Collective's activism focuses on a remembrance of this history, which is not part of the dominant narrative in the United States, as well as a reclamation of pre-colonial land and foodways and efforts toward food sovereignty (icollective2019_2020), defined by the U.S. Food Sovereignty Alliance during a 2007 conference as the right to healthy and culturally appropriate food, produced sustainably (Food Sovereignty 2019). The ways that colonisation has negatively affected food sovereignty, particularly for indigenous communities, remains a primary concern for American Indian activists today.

8.5 Power of Social Media

The I-Collective's Instagram feed, their main mode of social media connection, boasts more than 7,000 followers. Their posts include information on events hosted and attended by their members, typical "foodie"-like posts honouring native foods, and direct posts challenging colonialism. There is no doubt where this group stands politically, both challenging current policies and emphatic support of female and indigenous leaders. Despite—or perhaps because of—this pointed perspective, they were named Top Food Influencer by the popular publication *Uproxx* (Johnston 2019). Despite their (relatively) small reach compared to other top influencers, their

[2] This is also reflective of the female leadership roles practiced by most tribes: Traditionally, women were the craftspeople, the heads of household, and the tribe decision-makers.

approach to food and activism is intertwined in a way reminiscent of the Black Panther Party's Free Breakfast for Children Program. Their events feature lessons on indigenous history and oppression while serving food made only with pre-colonial ingredients, and their Instagram posts similarly work to inexorably link their food-ways and political ideals in what can be interpreted as a rhetorical homology. Below are two examples:

Example 1 Text (alongside an image of rack of drying meat):

> When your [sic] a Village Indian at heart, living in an urban environment, you do what you can to get that dry moose meat. This photo is courtesy of @3jenkins and her current moose meat drying process in California. When you live in urban places and aren't on your traditional lands, you have to improvise. This speaks to the resiliency that we have as Indigenous people to keep fighting for that traditional knowledge colonizers work so hard to take away. Tell us what some of your Urban Indian improvised techniques #urbanindian #villageindian#moose #moosemeat #dried-moose#improvised #mentastalakevillage#alaska #california #dryingmeat#traditional #traditionalknowledge#foodsovereignty#traditionalfoodknowledge#traditionalfood (I-Collective 2019a).

Example 2: Text (alongside the video of a member chef cooking):

> As a collective, we love to cook for folks, but it's even better when cooking with folks. On Saturday night of the @4cornerscollaborative's symposium, we had the pleasure of not only continuing to cook with Josh Nez (@cook_nez), Gary, Kyle, and Dineh (@dineh_redstar), but most importantly we cooked with Rosa. For the last two years, Rosa has sought sanctuary from deportation in the Methodist church in Mancos. Her story is a complicated, yet common narrative, in this country and has for many in this community created the fire to fight against the tyranny of this current administration and ICE. To learn more, or to donate to the legal fund, visit https://rosabelongshere.org/ ♥ (I-Collective 2019b).

Both examples exhibit a clear connection between foodways and the fight against coloniser practices, past and/or present, while also asserting a clear political viewpoint and, in cases, a call to action. Foodways serves as the conduit, both for the text message as well as for the events or actions featured in the posts. With the I-Collective's popularity still growing—they almost doubled their followers from fall 2019 to spring 2020—their reach may still be smaller than the media generated by and about the Black Panther Party's food security efforts, or #tacotruck—but their approach to community activism is the same. They recognise the power of food to connect and convince, and the ways that it can represent ideas and ideals that are so much broader than what is on the plate. Similar to the Black Panthers, the I-Collective understands that in addition to their media communications, that the true changing of minds and modelling behaviour often must happen in person. Enhancing their written, media-based rhetorical approaches, their periodic events marry nutritious and culturally appropriate meals with pointed education about American colonization of Native American tribes, modelling this tension through the cultural implications of the food on the plate.

8.6 Conclusion

The Black Panther Party used their Free Breakfast for Children Program as a tool for political discourse by creating a program that both addressed the real public problem of hunger while also making a space and opportunity to teach and discuss the socio-political context underlying that hunger. The Free Breakfast program allowed the Black Panther Party to both demonstrate and convey their socialist approach to material support, wealth distribution, and people power. Their unintended consequence was that these activities compelled the mainstream mass media to reframe the narrative of Black Panther identity and efforts (though negatively affected by COINTELPRO's propaganda initiatives) to a much larger audience, garnering them greater public support, and eventually, pressuring the government into providing these same food programs on a much larger scale. Their community organising efforts, despite the FBI's illegal actions, still influenced major federal policies that improve the lives of Americans daily, in addition to many other benefits for the communities with which they worked.

In comparison, the I-Collective deploys media more directly and purposely to educate, organise, and activate through memes, bite-sized history lessons, and the emphasis of food as indigenous culture. In many ways not only have the nature of media and public discourse changed, but also the manner in which people interact with it. While the I-Collective's influence is still growing, they are using events in ways similar to the Black Panther Party, while also taking part in the discourses at the intersection of food and politics on social media.

Despite these differences, these two case studies—spanning time, geographic location, and political context, among other differences—share some similarity of approaches to the use of food as a political tool for education and change, both creating a rhetorical homology that marries food and political discourses. The Black Panther Party's Free Breakfast for Children Program, and its accompanying discourse, taught activists then as today about the power of food: as both a daily necessity and as cultural capital for an entry point to reinforce a political agenda that is about so much more than what is on the plate.

References

Belasco, Warren. 2006. *Appetite for Change: How Counterculture Took on the Food Industry*. Ithaca, NY: Cornell University Press.

Brummett, Barry. 2008. Rhetorical homologies in walter Benjamin, the ring and capital. *Rhetoric Society Quarterly* 36: 449.

Carstairs, Philip. 2017. Soup and reform: improving the poor and reforming immigrants through soup kitchens 1870–1910. *International Journal of Historical Archaeology* 21 (4): 901–936. https://doi-org.proxy.library.nyu.edu/https://doi.org/10.1007/s10761-017-0403-8.

Churchill, Ward, and Jim Vander Wall. 1990. *The COINTELPRO Papers: Documents from the FBI's Secret War Against Dissent in the United States*. Boston: South End Press.

Cope, Suzanne. 2018. Justice among the Jell-O recipes: feminist history of food journalism. *Los Angeles Review of Books*. 9 July. https://lareviewofbooks.org/article/justice-among-the-jell-o-recipes-the-feminist-history-of-food-journalism/. Accessed 4 Aug 2020.

Cope, Suzanne. 2014. *Small Batch: Pickles, Cheese, Chocolate, Spirits, and the Return of Artisanal Food*. New York: Rowman & Littlefield.

Echo-Hawk, Hillel. 2018. Personal Interview. March 4.

Food Sovereignty. 2019. U.S. Food Sovereignty Alliance. http://usfoodsovereigntyalliance.org/what-is-food-sovereignty/. Accessed 23 Oct 2019.

Forsey, Caroline. 2019. The ultimate list of instagram influencers. *Hubspot*. October 30. https://blog.hubspot.com/marketing/instagram-influencers. Accessed 15 May 2020.

Heynen, Nik. 2009. Bending the bars of empire from every ghetto for survival: the black panther party's radical antihunger politics of social reproduction and scale. *Annals of the Association of American Geographers* 99 (2): 406–422. https://doi.org/10.1080/00045600802683767.

I-Collective (icollective2019_). 2019a. *When your a Village Indian at heart.... Instagram*. September 22. https://www.instagram.com/p/B2ujoSdlxqt/. Accessed 15 May 2020.

I-Collective (icollective2019_). 2019b. *As a Collective, We Love to Cook for Folks.... Instagram*. June 22. https://www.instagram.com/p/BzA5O7oloSY/. Accessed 15 May 2020.

Is It Too Late For You To Be Pals With a Black Panther? 1970. *Esquire* 74 (5): 141–147.

Johnston, Zach. 2019. Your instagram feed needs these "food influencers" In *The Mix ASAP, Uproxx*. July 24. https://uproxx.com/life/best-food-influencer-on-instagram-2019/. Accessed 15 May 2020.

Lopez, German. 2016. Trump Surrogate Warns of Scary Future with "taco trucks" on "every corner", *Vox*. September 2. https://www.vox.com/2016/9/2/12768774/trump-taco-trucks. Accessed 15 May 2020.

Murch, Donna. 2010. *Living for the City: Migration, Education, and the Rise of the Black Panther Party in Oakland*. Chapel Hill: University of North Carolina Press.

Perlman, Alison. 2013. *Political Appetites: Food as Rhetoric in American Politics*. Publicly Accessible Penn Dissertations 907. http://repository.upenn.edu/edissertations/907. Accessed 15 May 2020.

Pollan, Michael. 2017. Food and more: expanding the movement for the trump era. *Civil Eats*. January 16. https://civileats.com/2017/01/16/food-and-more-expanding-the-movement-for-the-trump-era/. Accessed 4 Aug 2020.

Potorti, Mary. 2014. Feeding revolution: the black panther party and the politics of food. *Radical Teacher* 98 (Winter 2014). https://pdfs.semanticscholar.org/1867/3443a305b77300ffd3a6cbf6f87a7df6b82f.pdf. Accessed 15 May 2020.

Popular hashtags for tacotruckoneverycorner on Twitter and Instagram. 2020. https://ritetag.com. Accessed 12 May 2020.

Radosh, Ronald. 2011. The politics of food. *Humanities* 32 (2). https://www.neh.gov/humanities/2011/marchapril/feature/the-politics-food. Accessed 2 May 2020.

Ruis, Andrew R. 2015. "The penny lunch has spread faster than the measles": Children's health and the debate over school lunches in New York City, 1908–1930. *History of Education Quarterly* 55 (2): 190–217. https://doi-org.proxy.library.nyu.edu/https://doi.org/10.1111/hoeq.12113.

Scott, Erica. 2018. Personal Interview. Feb 27.

Shapiro, Laura. 1986. *Perfection Salad: Women and Cooking at the Turn of the Century*. New York: Farrar, Straus & Giroux.

The Panthers' Bite. 1968. *TIME Magazine* 92 (12): 29.

Top 100 Hashtags on Instagram. 2020. Top-Hashtags.com, May 5. https://top-hashtags.com/instagram/. Accessed 15 May 2020.

Thompson, John R. 2012. Food talk: Bridging power in a globalizing world. In *Routledge Studies in Rhetoric and Communication: The Rhetoric of Food: Discourse, Materiality, and Power*, ed. Frye Joshua and Michael Bruner, chapter 4. London, UK: Routledge.

USDA. 2018. A Short History of Snap. September 11. https://www.fns.usda.gov/snap/short-history-snap. Accessed 15 May 2020.

Voss, Kimberly. 2014. *The Food Section: Newspaper Women and the Culinary Community*. Rowman & Littlefield.

Wexler, Mark N., Judy Oberlander, and Arjun Shankar. 2017. The slow food movement: a "big tent" ideology. *Journal of Ideology* 37 (1): 1–34.

Suzanne Cope is the author of *SMALL BATCH: Pickles, Cheese, Chocolate, Spirits, and the Return of Artisanal Food* (Rowman & Littlefield 2014) and the upcoming book *POWER HUNGRY: Women of the Black Panther Party and Freedom Summer and Their Fight to Feed a Movement* (Chicago Review Press, 2021). She has written on food, culture, and politics in popular venues including the *New York Times, BBC, The Atlantic*, and has presented and published in academic forums including ASFS and Italian American Review, among others. Dr. Cope has a Ph.D. in Adult Learning and an MFA in creative nonfiction. She teaches writing at NYU. E-mail address: suzanne.cope@nyu.edu.

Chapter 9
"Superfine Quality, Absolute Purity, Daily Freshness": The Language of Advertising in United Cattle Products' Marketing of Tripe to British Workers in the 1920s and 1930s

David Michael Bell and Theresa Moran

Abstract For more than 100 years, tripe was one of the most popular foods in the Northwest of England. In 1920, independent tripe shops were amalgamated into the United Cattle Products (UCP). Using promotional advertising in local newspapers, this paper examines UCP's tripe marketing strategies. To counter claims that tripe was unclean, ads stressed tripe's attributes of purity, quality and freshness; nutrition; digestion; economy; timesaving; and taste. An examination of images reveals a message that tripe is cooked in middle-class homes by stay-at-home moms and enthusiastically eaten by white-collar fathers and children, so fulfilling working people's aspirational social needs for a happy home and family. UCP's advertising also connected tripe to higher order needs of self-esteem and actualisation by validating such moral qualities as perseverance and endurance, qualities especially espoused by people in the industrial Northwest as part of their work ethic. UCP finally closed in the 1990s. Ostensibly, its decline is linked to cheap industrial meats. But a deeper and more disturbing cause of tripe's decline may lie in what can be seen as the constant and arguably unethical undermining of a community's foodways by the prejudicial forces of disdain (social class stigmatisation), disgust (social control), and ridicule.

Keywords Tripe · Advertising · Disdain · Disgust · Ridicule

D. M. Bell (✉) · T. Moran
College of Arts and Sciences, Ohio University, Athens, USA
e-mail: belld@ohio.edu

T. Moran
e-mail: morant@ohio.edu

S. Stano and A. Bentley (eds.), *Food for Thought*, Numanities - Arts and Humanities in Progress 19, https://doi.org/10.1007/978-3-030-81115-0_9

9.1 Introduction

This is a story about English tripe and more specifically about the tripe eating region of the north-west of England which has Manchester as its centre. Manchester was the first city to experience industrialisation, based initially on textiles and mass urbanisation. This is a story about what the working people of Manchester ate at a time when meat was more than double the price of offal and at best only eaten as a Sunday roast or on the religious feast days of Easter and Christmas. This is the story of UCP or United Cattle Products, a curious history of a food chain which in its day has been compared to the success enjoyed by MacDonald's today. This is the story of a remarkable food entrepreneur: John Septimus Hill, who later became the Lord Mayor of Manchester, 1942–43, during which time he entertained King George and Queen Elizabeth at his official residence. Hill founded UCP in 1920 with a mission to feed the working people of the greater Manchester area. And this is the story of UCP's attempts through its advertising to defeat the forces of disdain, disgust and ridicule, which stigmatized the consumption of tripe. This was a battle that UCP ultimately lost with the rise of cheap industrial meat following the Second World War. UCP closed in 1999 and today, tripe, which was once eaten by millions in the Manchester area, hardly exists anymore in the region's foodways.[1]

Offal refers to the edible parts of the animal that are discarded or "fall away". It derives from Middle Dutch *afval* ("fall off), *af* ("off") + *val* ("fall"). It was first used to describe mainly the entrails, but later extended to include the head, tail, and internal organs such as the heart, and liver. In American English offal is referred to as organ meats or the euphemistic "variety" meats. A more recent definition is: "Cast off and maligned", which would distinguish tripe from *foie gras* (Barthouil 2017).

Tripe is the stomach lining of usually oxen—castrated adult male cattle. An 800 lb ox yields about 15lbs of tripe. Oxen have four stomachs and provide four main types of tripe. Flat or seam tripe comes from the first stomach or rumen. Honeycomb Tripe comes from the lower part of the second stomach or reticulum. This is the most tender and meatiest tripe. Book or bible or leaf tripe comes from the third stomach or omasum. And reed or black tripe, so rippled in texture that it is known as "fisarmonica" (accordion) in Florentine dialect, comes from the fourth stomach or abomasum. In England, tripe, which is naturally green or grey, is always washed and dressed (that is, the stomachs are cleaned and the fat trimmed off), and then bleached for up to two days. It is then boiled for several hours so that the product offered for sale is usually creamy-white and ready to eat.

English tripe dishes are very simple. The two main ways of serving tripe are hot with onions or as a cold dish with lettuce and tomatoes. One of this paper's authors is old enough to remember the local UCP shop in Manchester where he grew up. He ate honeycomb tripe in salad and recalls that the gustatory trick was to fill each honeycomb with malt vinegar. His mother preferred the more pungent black tripe. It can be said that English tripe preparations have nothing of the invention of other

[1] For a collection of memories, documents and images related to the United Cattle Products Ltd., see: http://www.unitedcattleproducts.co.uk/index.php. Accessed 17 May 2020.

countries' tripe cuisines, for example, Italian tripe dishes, which vary from city to city. In Treviso, it is served with parsley and chicken stock; in Turin, with wild mushrooms; in Genoa, with beans, peas, and potatoes, in Milan, with butter and asparagus, and rice; in Lucca with cinnamon and cloves; in Florence, with green sauce and grilled peppers and of course in Rome, *trippa alla Romana* is served with onions and tomatoes (Di Renzo 2010: 76).

9.2 United Cattle Products

John Septimus Hill (1874–1949) was a visionary food entrepreneur. He had started work in his brother's butchering business but soon left to start a new venture of tripe dressing, selling it in the streets of Manchester from a handcart. In 1920, under the name United Cattle Products, Hill amalgamated a group of independent tripe-dressers in the Northwest of England. By 1939 UCP operated 166 restaurants and retail establishments and 12 factories and achieved record sales and profits (U.C.P Record Profits 1939).

Hill modelled his UCP shops on such meat chains as the Vestey Brothers' Dewhurst butcher shops, which were a regular feature of any English high street. Technological developments in refrigeration and steam ships enabled the import of foreign meats into the UK from North America, South America and later Australia and New Zealand. In 1880 meat imports accounted for just one sixth of meat consumption. That figure rose to one half by 1902 and by the outbreak of the first world war in 1914, the bulk of meat consumption in England was imported (Burnett 2013). By 1890, meat importers such as James Nelson and Sons Ltd, Eastman's Ltd, the River Plate Fresh Meat Co Ltd, W and R. Fletcher Ltd and London Central Meat Co Ltd had begun to develop chains of retailing frozen meat to the urban population, mainly in industrial midlands and the north of England (Oddy 2003: 18). By 1925 Vestey Brothers was the largest meat retailer in the world. The company controlled one third of the refrigerated storage capacity in Britain and 2,365 butchers' shops, or two thirds of the butchers' shops under the name of Dewhurst (Farrell 2015).

Although the British had more access to affordable meat than most other people in Europe, it was still too expensive for most working people. At best they ate meat in the form of a joint once a week on Sunday. But there were other issues apart from the cost of the meat itself which made meat a less viable source of nutrition for the urban working class. Living conditions were over-crowded and ill-equipped for cooking. The move from rural to urban living with restrictions on kitchens, time and products meant that a lot of culinary knowledge was lost (Broomfield 2007). The cost of fuel was another significant constraint on cooking meat during the week. Fuel costs meant that only two or three hot meals could be prepared per week. Many women, whose traditional role in the home was still that of cook and cleaner, also worked long hours in the factories, and had little time or energy for cooking during the week. Cooking was left to Sunday when there was enough prep and cooking time. Daily urban diets, therefore, meant greater dependence on ready-made foods like bread and those which needed the least preparation. Thus boiled and roasted potatoes, and bacon, which could be fried in minutes, became mainstays of urban diets (Burnett 2013).

John Septimus Hill understood that cooked food had made the Northwest the workshop of Britain, so he could see how tripe and the other offal products of the UCP were "the ideal food for Lancashire workers" (*Silver Jubilee of UCP* 1945: 8). Hill identified the consumers of UCP products, therefore, as "the teeming home-loving millions of the North, who need ready prepared appetizing food without trouble to cook and serve—those workers who ask for maximum of good food at minimum of cost" (Ibid.: 3).

9.3 Prejudice

Persuading people to purchase products necessarily involves an attempt to exploit sets of shared beliefs, values, attitudes, and biases, as well as the social identities connected to those products. Hill, like all good salespersons, was essentially an applied cultural anthropologist with regard to his knowledge of his customers' shared beliefs about work, life, and food. But he was also aware of others' beliefs that tripe and the consumption of tripe was considered unsavoury. Hill labelled these negative ascriptions of tripe as "prejudice" and believed that "any Pecksniffian prejudice arose through the misuse of the word itself or because tripe was sold in a careless unhygienic fashion" (The 99 Recipe Book 1932: 2).

According to the OED, the use of the word tripe was used figuratively more frequently after the 1890s to mean worthless or rubbish in reference to opinions, conversations, and especially in artistic works such as books, songs and later movies. In ads and in brochures, Hill sought testimony as to tripe's worth, quoting from Chaucer, Shakespeare, Goldsmith, Peyps, Burns, and Dickens (Tripe in Literature 1923). A brochure and subsequent ads entitled "Fit For a King" listed ancient Roman nobility and kings who ate tripe. Even King Edward VII was mentioned as being fond of tripe. In this way, UCP claimed to have "resuscitated its ancient fame" (The 99 Recipe Book 1932: 2).

Hill countered the claim that linked tripe to a lack of hygiene by elevating hygiene and cleanliness in his tripe works, shops and restaurants. For years prior to UCP, the typical tripe shop had a few small tables and chairs or a small room behind the shop called a Supper Bar where workers could eat a meal. UCP changed all that by establishing upscale tripe restaurants that combined delicatessens in the modernist style with art-deco interiors aimed at more sophisticated diners. The UCP Ralph Mason Luxe Cafe in Blackpool even employed a pianist. (Evidence of Accused 1921). By 1945, UCP had established 37 tripe restaurants and there was a UCP restaurant in every major town in the northwest. In the centre of Manchester, UCP had three restaurants. Hill's UCP Restaurant on Market Street had a lift to all floors and a main dining room that seated 300 with "vita-glass windows [that] allow[ed]unhindered sunshine to bathe [the] interior with ultra-violet rays of tonic quality" (*Silver Jubilee of UCP* 1945: 40). The restaurant also boasted an American Bar and a Cloak Room and Ladies Retiring Room "with attendants embodying those refinements that mean so much to our patrons after shopping, and before going to a Show, a Concert or a Cinema" (Ibid.: 40). The UCP restaurant on Oxford Street, in the heart of the theatre district, had timbered roofs and panelled walls in the style of the Tudor period (Ibid.: 51). Similarly, the description of the kitchens further belies the contemporary and modern stereotypes of an offal restaurant. The kitchens are described as the "last word in modern efficiency and hygiene" and as "models for kitchens of the future" with "all the most scientific methods of cooking and serving the finest foods … with scrupulous cleanliness everywhere, including air conditioning to maintain freshness and a cool atmosphere" (Ibid.: 40). UCP advertised its major restaurants as venues for "Banquets, Dances, Social Gatherings, Ladies' Evenings, Conferences, Weddings, Birthday Parties, and the like" (37). Hill could rightfully claim that UCP restaurants combined luxury with economy. In 1945, these restaurants offered a three-course meal for 1 shilling and 7d or £2.81 today adjusted for inflation. "UCP restaurants", boasted Hill, "lifted the standard of service and set a very high ideal in appearance, conduct, cleanliness, and courtesy" (Ibid.: 8).

9.4 Advertising Tripe

In 1922, UCP started their first advertising campaign, working with Edward R Cross of the Manchester based Cross-Courtney Company. First, UCP introduced the red oval shield which identified each of their shops. Each UCP shop carried "show-cards"—placards (see Fig. 9.1) and brochures. UCP shops were painted bright red; Hill wanted them to stand out as "bright spots on the drab streets" (*Silver Jubilee of UCP* 1945: 13) Hill's fleet of delivery vans and lorries were used as mobile hoardings (billboards) displaying colourful posters extolling the enjoyment of eating tripe. For the most part, print advertising was focused on ads in local newspapers throughout the Northwest. These ads included competitions for kids. They could win a bicycle or a toy airplane by drawing "the funniest ox and sheep and pig you know". They could also win a holiday by the sea writing a simple essay: "Why I Iike UCP tripe". Further, there were educational incentives for teachers and schools: for any child who won a bike the UCP promised a valuable prize of books to the child's school.

UCP also targeted the home cook with *The 99 Recipe Book*. First published in 1923, the book contains *Ninety-nine homely and delicious ways of preparing and serving UCP tripeand cowheels*. In 1935, UCP even used popular records to promote tripe. Stainless Stephen sang *Hot Tripe* on one of the first cardboard records (Houlihan 2011). The label reads: "Stainless Stephen is a hot Radio Favorite who sings the praises of the finest winter food in the world" (The Prime Minister of Punctuation 1935). Hill summed up the campaign: "In a thousand and one ways the virtues of this so delicious tripe, but so inexpensive and nutritious food, were hammered into the consciousness of Northern folk" (*Silver Jubilee of UCP* 1945: 13). By 1924,

Fig. 9.1 UCP showcard in the 1940s

Advertisers Weekly recognised UCP's efforts as "the most outstanding and the most courageous advertising campaign of the year" (Ibid.: 13).

To examine UCP's advertising in newspapers, we searched the British News Archive from 1920 onwards using the search terms 'tripe' and 'UCP.' We begin by analysing the textual thematic content, and then we look at the images in the ads using the concept of re-contextualisation. Finally, we examine the ads according to Maslow's (1943) hierarchy of needs.

9.4.1 Textual Thematic Content

We used thematic analysis to capture the major arguments used in the ads to sell tripe. We identified 6 major attributes of tripe that the UCP advertising campaign constantly stressed throughout the 1920s and 1930s: Purity, Quality and Freshness; Nutrition; Digestion; Economical; Timesaving; and Taste. Below we examine these attributes.

Freshness, Purity and Quality: Here are a few examples of the language the ads used to stress the freshness, purity and quality of its tripe:

- "Superfine quality, absolute purity, daily freshness";
- "An absolute guarantee of purity, quality, and daily freshness";
- "UCP Pure Food Products";

- "Prepared in model Cookery Establishments under perfect Hygienic Conditions";
- "Uniformly Concentrated Purity".

Hill was very proud of the modern hygienic conditions in his tripe works: "Each is maintained and operated to provide the utmost service in hygiene, in quality, and in the maximum of its pure food products" (*Silver Jubilee of UCP* 1945: 34). Hill insisted on his workers' frequent change of garments, overalls, and shoes. Hill even advertised tours of his tripe works for the public. His Tripe Works at Monton, Eccles was purposely built in 1938 with "garden city surroundings". UCP also had a fleet of 100 delivery vehicles that supplied the UCP shops and restaurants every morning. The copy in one ad read:

> U.C.P Tripe is guaranteed Fresh as Dawn. Early every morning, while the World and his wife are still abed, so scrupulously clean motors convey fresh supplies from the 14 U.C.P model cooking Establishments to all shops displaying the U.C.P. Oval Red Sign (As Fresh as the Dawn 1925).

Nutrition: UCP constantly stressed that tripe was as nutritious as more expensive cuts of meat and advertised it as a health food that created both mental and physical strength. Here are a few examples of the language the ads used to highlight the nutritional value of tripe:

- "Instantly nourishing";
- "In the interests of Health and Economy – eat more Tripe and Cowheels";
- "Its nourishing goodness brings health and happiness";
- "Tripe is as nutritious as red meat";
- "UCP Tripe is beyond question the most nutritious and most easily digested food. It is just the food for the sedentary worker. It is the creator of mental and physical strength".

These claims are substantially true. Like most organ meats, tripe is extremely nutritious. It is a good source of protein: 12grams per 100 gram (3.5 ounce) serving. Although this is half the protein found in an equivalent serving of beef steak, tripe is rich in B vitamins and minerals, including calcium, iron, magnesium, phosphorus, zinc, manganese and especially selenium, and tripe has less fat than an equivalent amount of beef steak (USDA National Nutrient Database 2020).

Digestion:

- "Easily digestible";
- "Mother knows that health waits on good digestion";
- "While Beef or Mutton requires 3½ hours for digestion UCP Tripe is completely digested within an hour";
- "Cooped up in a city. You can't keep going on if your digestive organs are clogged with heavy food. A meal of UCP Tripe is delicious and it leaves your brain clear for effort".

The claim that tripe is digested more than 3 times faster than beef was well documented. (See for example "Report on the best Means to Supply the Poor" 1842: 36).

Economical: One of the main benefits of eating tripe was that it was cheap compared with non-offal meat. Tripe actually stayed the same price: 9d (4p) a pound from 1923 until 1945. In addition, it was economical because it was pre-cooked and so saved on fuel costs.

- "Pre-War Food Prices At Last!" (April 1922);
- "Half the price of red meat" (January 1924);
- "Buy a pound and eat a pound. All food, no waste";
- "Saves fuel";
- "No coal! No coke! No cook!";
- "Cut Down Kitchen Time. And not only that, cut down meat and fuel bills as well".

Timesaving: Urban life, long working hours and minimal cooking facilities meant a greater dependence on readymade foods. Because tripe was already cooked, it required minimal time to prepare.

- "Easily prepared";
- "Quickly made";
- "Sold to you perfectly cooked, only requires heating";
- "It is but a matter of moments to serve it hot or cold";
- "The ever ready meal".

Taste: To the modern eater, the notion that tripe would be associated with the following purely evaluative adjectives seems incongruous:

- "Delicious";
- "Appetising";
- "Tasty";
- "Dainty UCP Tripe".

The use of "dainty" with tripe may seem especially contradictory, but "dainty" at one time had the meaning of a "choice or tasty morsel", which may, when referring to tripe, sound even more incongruous. *The 99 Recipe Book*, first published in 1923, and reprinted and expanded in 1932, was primarily an attempt to market tripe as a main ingredient of savoury dishes in the style of the Italian regional tripe dishes described earlier. *The 99 Recipe Book* contained such dishes as Tripe and Tomato Pie, Tripe Pasties, Honeycomb Soup, Tripe with Dumplings, Tripe and Beef Steak Pudding as well foreign influenced dishes such as Tripe a la Mode de Caen, Tripe a la Napolitaine, Tripe en Casserole, Gateau of Tripe, Curried Tripe, and Kedgeree of Tripe. But *The 99 Recipe Book* was also an attempt to unearth the recipes of mothers and grandmothers. The move from rural to urban living with restrictions on kitchens, time and products meant that a lot of culinary knowledge was lost and along with any number of dishes that eighteenth-century people of all classes consumed (Broomfield 2007).

9.4.2 Images and the Concept of Re-contextualisation

Now we want to switch from an analysis of the linguistic content of the ads to the non-verbal visual elements of the ads. We use here the framework of re-contextualisation (Van Leeuwen and Wodak 1999; Aboussnouga and Machin 2013) in order to examine the role that images play in UCP advertising. Re-contextualisation refers to the way that pictorial or textual representations of concrete participants, processes, and settings are re-contextualised through a process of deletion, addition, and substitution, which ultimately constitute an evaluation of that which is re-contextualised.

We use Leonardo da Vinci's *The Last Supper* or *Il Cenacolo* in Italian (Fig. 9.2) as an example of the re-contextualisation of an historical event. A representation cannot capture all the aspects of an event or social practice so it is pertinent to examine which participants, actions, and settings have been deleted. With regard to Da Vinci's depiction of the "Last Supper" nobody is sitting on one side of the table. It gives the impression that the setting is that of a high table at a banquet (Reinhartz n.d.) as distinct from the modest private dinner as seems to be the case in the gospels. In terms of addition, there is the use of a white linen tablecloth, typical of Italian textile production in the Renaissance, and various other tableware, which would not have existed at the original Last Supper (Varriano 2008). There is also the formal Renaissance architecture and tapestry wall hangings. With regard to substitution, we are thankful to the cleanup and restoration of the painting in 1999 for revealing just what food is on the diners' plates. They are apparently eating the very non-kosher roasted eels with orange slices, not the simple meal of dried figs, dates and bitter herbs most likely served at the original supper (Feeley-Harnik 1994). In terms of evaluation we can say that these diners are a special elevated group of people—many wear blue, a sign of wealth—and are eating in a fine palazzo. Da Vinci's reimagining of the Last

Fig. 9.2 Leonardo da Vinci, *The Last Supper* (1494–1498), Fresco, S. Maria delle Grazie, Milan

Supper represents, then, quite a change from the original simple group of disciples and their leader.

Applying this framework to the images in the UCP ads (see Fig. 9.3), we find that with regard to deletion there appear to be no working women. Women are represented as housewives, waiting at home for the husband's return from work and standing ready to serve up a hot meal. In the 1930s, more than one third of all women worked nationally in Britain, but this figure would have been much higher in the Manchester conurbation. Holloway (2005) puts the percentage of married female workers nationally at 16% in the 1930s. But again, this figure was no doubt higher in industrialised areas like Manchester amongst the working classes and especially in such industries as textiles which often employed as many women as men (Burnette 2008). With regard to addition, we can see that children appear frequently in UCP ads as enthusiastic consumers of tripe. Although this author relished tripe as a child, he cannot say the same for his siblings. UCP's junior tripe eaters would appear in reality to be aspirational. As for substitution, a middle-class white-collar father has been substituted for a working-class father. The father is always depicted as wearing a three-piece suit with a white shirt and tie. He and his wife and children suggest a family that could easily afford meat but here they are eating tripe. Thus the UCP ad images convey the impression that tripe is cooked in middle-class homes by stay-at-home moms and enthusiastically eaten by white-collar fathers and children. There seems to be a contradiction, then, between the identified consumer of tripe as described in UCP's mission: the working-class who have little time to cook, and the idealized consumer of the ads. The depiction of the idealised middle-class family eating tripe may be an attempt to legitimize tripe in the face of the threats of disdain, disgust, and ridicule.

Fig. 9.3 UCP print advertisement 1932

9.4.3 *Advertising Tripeand the Hierarchy of Needs*

UCP's combination of text and image represents a move from the purely informational style of advertising to the aspirational, a characteristic development in advertising in the early decades of the twentieth century. The idea was to encourage customers to associate a product with some aspect of lifestyle, personal improvement, or significant life event (Danesi 2015: 4). Using Maslow's (1943) hierarchy of needs we can see how UCP's advertising transitioned from emphasizing how tripe fulfils physiological needs: purity, nutrition, digestibility, economy, and taste, to fulfilling social needs: happy homes and family, and even the higher order needs of esteem and self-actualisation. This following ad was part of a series which connected the eating of tripe to the attainment and validation of universal moral qualities such as perseverance, endurance, and resistance, qualities that the industrial workers of the Manchester area would have espoused as part of their work ethic. Each ad featured an illustration and a parable. The following ad is captioned: PERSEVERANCE, and has an illustration of a navvy or a manual labourer driving a spike into concrete. The text is:

> At the twentieth stroke of the hammer the steely concrete split.
>
> "My word," said an onlooker, "It took that last blow to smash it, didn't it?"
>
> The navvy wiped his brow.
>
> "No, *it took every one of those twenty hits!"*
>
> The secret of perseverance!
>
> Perseverance is faith in the cumulative value of effort.
>
> (Perseverance 1936)

By 1932, UCP could claim that "the Prestige of Tripe has gone up by leaps and bounds" (The 99 Recipe Book 1932: 2) and that even "Prejudice has been replaced by Prestige" (Ibid.: 3).

9.5 Disdain, Disgust, and Ridicule

Yet, despite the success of UCP's "rebranding" and recontextualisation of tripe through its advertising campaign, the 'gentrification' of UCP's shops, restaurants, and production facilities, and Hill's tireless boosterism, a shareholder at the annual meeting in 1935 could still express her fears that tripe was still looked upon in a derogatory manner. "Is modern taste altering", she asked, "or is it the taste of tripe?" (Unfair To Tripe 1935).

The derogation of tripe that the UCP shareholder was referring to is vividly captured in George Orwell's bleak description of the living and working conditions among the working class in the industrial north of England. Published in 1937, *The Road to Wigan Pier* begins with a depiction of Orwell's sordid lodgings above an "ordinary 'tripe and pea' shop" and his lurid description of the Bookers, the owners of the shop and lodging house.

> Mr. Brooker was a dark, small-boned, sour, Irish-looking man, and astonishingly dirty. I don't think I ever once saw his hands clean. As Mrs. Brooker was now an invalid he prepared most of the food, and like all people with permanently dirty hands he had a peculiarly intimate, lingering manner of handling things. If he gave you a slice of bread-and-butter there was always a black thumb-print on it. Even in the early morning when he descended into the mysterious den behind Mrs. Brooker's sofa and fished out the tripe, his hands were already black. I heard dreadful stories from the other lodgers about the place where the tripe was kept. Black-beetles were said to swarm there. I do not know how often fresh consignments of tripe were ordered, but it was at long intervals, for Mrs. Brooker used to date events by it. 'Let me see now, I've had in three lots of froze (frozen tripe) since that happened,' etc. etc. We lodgers were never given tripe to eat. At the time I imagined that this was because tripe was too expensive; I have since thought that it was merely because we knew too much about it. The Brookers never ate tripe themselves, I noticed. (8)

One wonders what kind of advertising could ever seek to defeat such condemnations, such kind of 'othering' of a foodstuff. As Sharrock (2011) suggests, such descriptions "makes great copy" but bears little resemblance to reality. It is generally believed that Orwell moved to live with the Forrests, the real name of the Brookers, because of their low reputation and that his previous lodgings were too clean. For Orwell, the Brookers' tripe shop fitted perfectly into his southern prejudicial view of northern working class foodways. Yet at the same time, Wigan could boast the Tripe De Luxe Restaurant and Tea Room, opened in 1917, seating 300 diners with panelled walls, palm trees and a ladies' orchestra (Shipperbottom 1995: 2011).

Hill must surely have understood that what he referred to as prejudice was much more than the misuse of a word or concerns about hygiene, which he believed could be remedied by a clever advertising campaign and upmarket restaurants. The prejudice that Hill had to deal with, we suggest, can best be understood as an irresistible combination of disdain, disgust, and ridicule; disdain of a food considered low class, the psycho-social disgust of offal as impure, and ridicule in national popular media.

Hill would have understood quite clearly Bourdieu's (1986) concept of how social class membership can be manifested through the objectified cultural capital of food choices. This concept is exemplified in the notion of the *quinto quarto*. In Testaccio, the district of Rome which housed its slaughterhouses into the 1970s, the practice was to divide cattle carcasses into quarters for distribution according to social class rank. The prime cuts went to the aristocracy; the second-best cuts to the clergy; the third to the middle classes; and the fourth to the military. What was left—the offal, was called the quinto quarto or 'fifth quarter', and went to the lower classes (Edwards 2013). So quite clearly, the eating of tripe is connected to lower social class membership. An upward mobility would be marked by a change of eating habits from tripe to meat.

Disgust in its purest sense is a system in the brain that helps humans through smell and taste to avoid disease and contamination. The verb *to disgust* is a combination of the negative prefix dis- and *gust* meaning "taste", derives from the Old French *desgouster*, which in turn is believed to be derived from the Italian *disgustare,* meaning "to distaste" or in the adjective form "bad taste". Disgust is recognized as a basic emotion with a characteristic facial expression—closing the nostrils to shut off smell and opening of the mouth or gaping to allow contents of the mouth to dribble out, a distinctive physiological manifestation—nausea, a feeling of revulsion, and a physical distancing reaction of the body (Rozin and Fallon 1987). Rozin and Fallon define disgust specifically as a food related emotion: "a form of food rejection which is characterized by revulsion at the prospect of oral incorporation of an offensive and contaminating object" (Ibid.: 24). They posit three possible motivations for food rejection: (i) bad taste or odor; (ii) anticipation of harm following ingestion, either bodily in the short term as stomach cramps or long term in the form of cancer, or socially such as loss of social status through eating or being associated with "lower-class food"; and (iii) knowledge of the origin or nature of the food (24). Most food considered disgusting are animals, parts of animals, especially those that more closely resemble human parts, and animal by-products. Here is a typical example of disgust for offal from a food-related blog:

> I won't mince words here. The thought of eating offal–i.e. organ meats and other parts of animals–makes me want to hurl and/or run for the nearest exit. I won't tell you that brains, tripe, eyeballs, hearts or blood in general don't taste good. It's just that I don't want to try them to find out. I start gagging as the thought of something like that crossing my lips seems just wrong. It's a mental thing with me. I can't get past what I would be eating. (Farley 2012).

Despite conceding the potential good taste of offal, the rejection consists of the anticipated bodily harm and the origin of the food.

An interesting example of the possible fear of loss of social status through eating or associating with "lower-class food" is the derivation of the Italian word *schifo* as used in the phrase *Che schifo*—"How disgusting!" from "uno schifo", a skiff-shaped tray, which tripe sellers or *Tripparoli*, carried through the streets of cities like Rome. In addition to tripe, the *schifo* would contain pig's feet, and calf's heads (Di Renzo 2010: 78). Exclamations of *Che schifo!* may serve as announcements of class membership or aspiration. Likewise, exclamations of "How disgusting!" with the accompanying facial expressions serve as a powerful means of social control in situations like schools, where individual student food choices in packed school lunches may be severely restricted in order to maintain group identity.

The most interesting example of the ridiculing of tripe has been *The Fosdyke Saga*, a British comic strip by cartoonist Bill Tidy (1972–1985), published in the *Daily Mirror* newspaper from March 1971—February 1985. *It was later* adapted as a TV series 1977–83, a radio serial, and a stage play. The comic strip was a parody of John Galsworthy's classic novel series *The Forsythe Saga. The Fosdyke Saga* was the story of Roger Ditchley, the wastrel son of tripe magnate, Old Ben Ditchley, no doubt modeled on John Septimus Hill. The irony here is that the cartoonist Bill Tidy was deeply proud of his working-class roots in the Northwest of England. It certainly was not his intention to ridicule northern foodways, but the effect of his strip was to connect tripe eating quite clearly with clogs and shawls—the traditional clothes of the mill worker as depicted in the industrialscapes of Manchester artist L.S. Lowry, and other artefacts of a nostalgic past, which had no part in a modern progressive vision of the Northwest.

9.6 The Decline of Tripe

Although UCP continued to trade profitably in the post-Second World War period—profits were £209,000 in 1968 (£1,576,657.50 adjusted for inflation in 2020) and shareholders received a 15% dividend—the company declined rapidly in the 1970s and 80s, hanging on until the 1990s when their last shops and restaurants closed. Today, there are very few people in the Northwest who eat tripe. Although tripe briefly became fashionable again with gourmands in the 2000s due in part to Gordon Ramsay featuring it on one of his cooking shows (How Gordon Ramsay made a star out of tripe 2007) and the nose-to-tail trend in eating (Strong 2006), tripe never again was part of the foodways of millions of northerners, although the Northwest remains the home of black pudding, a sausage made with pig's blood.

The ostensible reasons for the decline of tripe was the greater affluence in post-war Britain which meant that working people could afford to eat more desirable cuts of meat. At the same time the rise of cheap industrial meats and the rise of "big chicken" made meat readily affordable for all (McKenna 2017). A roast chicken, which the working classes had only eaten at Christmas, now became a regular weekly meal. At the same time, however, tripe dressers were being forced out of business by the need for expensive equipment to maintain new stricter hygiene regulations which reduced their profit margins (Shipperbottom 1995: 2011).

But the question remains: Why would such a popular regional dish virtually disappear? After all tripe eating continues to be integral to foodways in other parts of the world (Edwards 2013). It is true though that, despite *The 99 Recipe Book*, English tripe cuisine never reached the tastiness of Italian, Spanish, and Portuguese tripe dishes. It is also true that tripe has never been able to package itself in a similar way to chicken, using such appealing and de-animalising language as nuggets, tenders, Buffalo wings, and drumsticks.

A further explanation is that the northern workers were tripe eaters by default, and that poor Britons would revert to the cultural stereotype of roast beef eaters once it was affordable. The iconic representation of the superiority of the English roast beef eaters or *Les rosbifs* is captured in Hogarth's (1748) painting *O, the Roast Beef of Old England* (or *The Gate of Calais*). The title was taken from a popular tune of the day, which extolled roast beef as the symbol of Britain's wealth and power. In the centre of the picture, a huge side of British beef is being carried to the English Inn at Calais, which is in stark contrast to the potage that the French are eating. This notion of reversion to cultural stereotype is analogous with the U.S. government's attempt to encourage Americans during the Second World War to eat offal or what was euphemistically called "variety" meats, because pork and beef were being shipped overseas to feed American soldiers. Once the war was over, Americans very quickly reverted to meat. As Wansink (2002: 93) noted, organ or variety meats were what "patriots" ate, not necessarily poor people. Rozin and Fallon (1987: 27) note that because most things disgusting like tripe are nutritious, nutritional needs may offset feelings of disgust until such times that non-disgusting nutritional food is available again.

There certainly is a way that Hill thought of the northern working-class as "patriots" in the way they created wealth though their labour fuelled by the ideal food for working people. But whereas the United States had not really had a tradition of offal eating as each wave of immigrants quickly threw off their impoverished cuisines from the countries they left to mark their new prosperity by the eating of more expensive (non-offal) meat, the northern industrial workers had a long established offal foodways. The answer must be, therefore, that consumers who could now afford more expensive cuts of meat, eventually gave into the irresistible stigmatisation of tripe through a complex combination of disdain, disgust, and ridicule, the very forces that UCP fought so hard against through its marketing of tripe.

One modern powerful example of the ridiculing of tripe is the spoof *Tripe MarketingBoard* (2020), a website which parodies defunct British government organizations such as the Milk and Potato Marketing Boards. Here is an example of humour from the "recipe" section of its website: "Many people tell us they are put off trying tripe, but there are plenty of new tripe recipes available which successfully mask either its taste, smell, appearance or texture". One of the recipes is: "Pesto alla wiganese—The perfect fusion of Lancashire and the Mediterranean". The recipe combines tripe, Lancashire cheese, mushy peas, olive oil and pine nuts into a pesto. It is amusing in a similar way to stand-up comedian Jim Gaffigan's riffs on Hot Pockets (Merwin 2014).

But ridicule combined with disdain and disgust can be a powerful social control on a community's foodways. It also raises larger ethical questions. Could a food lobby deliberately target a key element of a group's foodways to ensure its collapse and so remove a competitor from the market? Given the erosion of consuming the once popular offal and the consequent food waste of eating only selected parts of an animal, can the forces of disdain, disgust and ridicule be seen as contributing to unsustainable food systems? Are these forces tools for manipulating individual food choices, such that it is hard to believe that most individuals would not easily choose to eat tripe?

9.7 Conclusion

Tripe had been a part of British foodways for centuries. With the industrial revolution and urbanization, tripe consumption became centred on the industrial Northwest of England. Tripe provided an economical, ready-to-eat, nutritious foodstuff for workers in the mills and factories. John Septimus Hill amalgamated many of the independent tripe dressers into a company which branded tripe under the name United Cattle Products (UPC) in its shops, restaurants, and factories. Through its advertising, UCP sought to legitimise, and elevate tripe eating as an integral and entrenched part of the regional foodways and social identities of millions of workers in the Northwest. Ostensibly, UCP's advertising campaign addressed the prejudicial views of tripe as worthless and unclean by stressing UCP tripe's attributes of purity, quality, freshness, nutrition, digestion, price, timesaving, and taste. At the same time, UCP advertising confronted those larger and more complex prejudicial forces, of what we have identified as disdain, disgust, and ridicule, by promoting tripe eating as fulfilling the aspirational social needs of a happy home and family. UCP's advertising went further by connecting the eating of tripe to attaining higher order needs of self-esteem and actualisation through the validation of such universal moral qualities as perseverance and endurance, qualities of a work ethic especially espoused by the industrial workers of the Northwest. Ultimately, tripe, central to the foodways of millions, disappeared almost completely in just 30 years. The rise of cheap industrial meat allowed the forces of disdain expressed through social class stigma, the social control of disgust, and ridicule to erase the eating of tripe from the social identities of the people of the Northwest. Consumers eventually gave into the irresistible stigmatisation of tripe, the very forces that UCP's marketing fought so hard against.

The example of tripe and its erasure begs the question of whether social control exerted over people's food choices will undermine the sustainability of our food system. Given the erosion of a popular offal foodways—efficient and nutritious—and the consequent food waste from eating only certain parts of an animal, can disdain, disgust, and ridicule, be called out as highly unethical responses to food choices?

References

Aboussnouga, Gill, and David Machin. 2013. *The Language of War Monuments*. London: Bloomsbury Academic.

As Fresh as the Dawn. 1925, January 10. *Burnley Express*, p. 13. col. 5.

Barthouil, Guillemette. 2017. Foie gras: the quantum offal In *Offal: Rejected and Reclaimed Food: Proceedings of the Oxford Symposium on Food and Cookery 2016*, ed. Mark Mcwilliams, 72–84. Totnes, UK: Prospect Books.

Broomfield, Andrea. 2007. *Food and Cooking in Victorian England: A History. Victorian* Life and *Times* Series. Westport, CT: Praeger.

Bourdieu, Pierre. 1986. The forms of capital. In *Handbook of Theory and Research for the Sociology of Education*, ed. J. Richardson, 241–258. New York: Greenwood.

Burnett, John. 2013. *Plenty and Want: A Social History of Food in England from 1815 to the Present Day*, 3rd ed. London: Routledge.

Burnette, Joyce. 2008. *Gender, Work and Wages in Industrial Revolution Britain*. Cambridge: Cambridge University Press.

Danesi, Marcel. 2015. Advertising Discourse. In *The International Encyclopedia of Language and Social Interaction*, ed. Karen Tracy, Cornelia Ilie, and Todd Sandel, 1–10. New York: Wiley.

Di Renzo, Anthony. 2010. *Greens: Essays on Food, Politics, and Ethnicity from the Imperial Kitchen: Essays on Food, Politics, and Ethnicity from the Imperial Kitchen*. Albany: State University of New York Press.

Edwards, Nina. 2013. *Offal: A Global History*. London: Reaktion Books.

Evidence of Accused. 1921, May 12. *Lancashire Evening Post*. p. 2 para. 3.
Farley, Mike. 2012. Offal = Awful. *Mikey's Kitchen*. https://www.mikeyskitchen.com/2012/10/27/offal-awful/. Accessed 11 May 2020.
Farrell, Thomas. 2015. Meat The House of Vestey. http://letslookagain.com/2015/04/invested-a-history-of-vestey-brothers/. Accessed 14 May 2020.
Feeley-Harnik, Gillian. 1994. *The Lord's Table: The Meaning of Food in Early Judaism and Christianity*. Washington: Smithsonian Institution Press.
Hogarth, William. 1748. *O the Roast Beef of Old England ('The Gate of Calais')*. Tate. https://www.tate.org.uk/art/artworks/hogarth-o-the-roast-beef-of-old-england-the-gate-of-calais-n01464. Accessed 29 Apr 2020.
Holloway, Gerry. 2005. *Women and Work in Britain since 1840*. London: Routledge.
Houlihan, Marjory. 2011. *Tripe: A Most Excellent Dish*. Totnes, UK: Prospect Books.
How Gordon Ramsay made a star out of tripe. 2007. Thursday 12 July. *Evening Standard*. https://www.standard.co.uk/showbiz/how-gordon-ramsay-made-a-star-out-of-tripe-7265246.html. Accessed 2 May 2020.
Maslow, A.H. 1943. A theory of human motivation. *Psychological Review* 50 (4): 370–396.
McKenna, Maryn. 2017. *Big Chicken: The Incredible Story of How Antibiotics Created Modern Agriculture and Changed the Way the World Eats*. Washington, DC: National Geographic.
Merwin, Hugh. 2014. Jim Gaffigan May have killed hot pockets once and for all. *Grub Street*, October 23. https://www.grubstreet.com/2014/10/hot-pockets-sales-are-declining.html. Accessed 10 May 2020.
Oddy, Derek J. 2003. *From Plain Fare to Fusion Food: British Diet from the 1890s to the 1990s*. Woodbridge, Suffolk: Boydell Press.
Orwell, George. 1937. *The Road to Wigan Pier*. New York: Houghton Mifflin.
Perseverance. 1936, January 22. *Lancashire Evening Post*, p. 2, col. 6.
Reinhartz, Adam. n.d. The last supper. Focus on last supper. *Oxford Biblical Studies Online*. https://global.oup.com/obso/focus/focus_on_lastsupper1/. Accessed 3 May 2020.
Report on the best Means of Supplying the Poor Cheap and Nutritious Food. 1842. *Proceedings of the Royal Philosophical Society of Glasgow, Volume 1 Glasgow 1841–42*: 29–40. Richard Griffin & Co. Reprinted by Pranava Books, 2019.
Rozin, Paul, and April E. Fallon. 1987. A perspective on disgust. *Psychological Review* 94 (1): 23–41.
Sharrock, David. 2011. The road to wigan pier, 75 years on. *Guardian*. Saturday, February 19, para. 11. Accessed 27 April 2020.
Shipperbottom, Roy. 1995. The decline of tripe. In *Tripe a Most Excellent Dish*, ed. Marjory Houlihan, 2011. Totnes, UK: Prospect Books.
Silver Jubilee of UCP: The story of a great FOOD INDUSTRY just a quarter century young 1920–1945. 1945. United Cattle Products Limited. Manchester, England.
Strong, Jeremy. 2006. The modern offal eaters. *Gastronomica* 6 (2): 30–39.
The 99 Recipe Book. 1932. United Cattle Products Ltd: Manchester. Reprinted 2013 by Aerostato: Seattle.
The Prime Minister of Punctuation: Stainless Stephen Makes Gramophone Record. 1935, October 26. *Burnley Express*, p. 19. col.5.
Tidy, B. 1972–1985. *The Fosdyke Saga*. vols 1–14. Daily Mirror Newspapers.
Tripe in Literature. 1923, October 27. *Burnley News*, p. 9, col. 6.
Tripe Marketing Board. 2020. https://tripemarketingboard.co.uk. Accessed 1 May 2020.
UCP Record Profits. 1939, May 4. *Manchester Evening News*, p.8. col. 1.
Unfair To Tripe. 1935, May 17. *Yorkshire Evening Post*, p. 10, col. 5.
USDA National Nutrient Database. (2020). https://fdc.nal.usda.gov/fdc-app.html#/food-details/174769/nutrients. Accessed 3 May 2020.
Van Leeuwen, Theo, and Ruth Wodak. 1999. Legitimizing Immigration: a Discourse-historical Analysis, *Discourse Studies* 1 (1): 83–119.
Varriano, John. 2008. At supper with leonardo. *Gastronomica* 8 (1): 75–79.

Wansink, Brian. 2002. Changing eating habits on the home front: Lost lessons from world war II research. *Journal of Public Policy and Marketing* 21 (1): 90–99.

David Bell is Associate Professor and Chair of Linguistics at Ohio University. His research focus is on Semantics and Pragmatics, and Language, Culture and Food.

Theresa Moran launched the Food Studies Theme at Ohio University in 2013. She is the University's Sustainable Living Hub Coordinator and Adjunct Associate Professor of Instruction in Environmental and Plant Biology.

Chapter 10
New Generations and Axiologies of Food in Cinema and New Media

Bruno Surace

Abstract Until recently the foodsphere was an exclusive domain of the adult world. Except in rare cases, parents' or caretakers' beliefs largely determined children's and adolescents' nutrition. It was with the Millennials, and later with the so-called Generation Z, that the theme of food first gained prominence in the media, with the multiplication of texts on the subject, and with progressive attention being paid to environmental crises, began to involve young people. Today, food is one of the favourite amusements of younger generations, who enjoy cooking, shopping, watching movies, documentaries, online videos where the most disparate value systems are articulated, framing food inside a discourse regarding its nutritional and ethical value, and on the systems of production that underlie it. Food is now far from being mere nutrition for young people, but more and more a linguistic element (think of the metaphor of "binge watching"), an identity, a community, something aesthetic (in various senses), valuable, semiotic. The aim of this paper is to investigate this vast territory, trying to map and shape the main axiologies (or value assets) in the postmodern foodsphere of and for the new generations.

Keywords Food and media · Food cultures · New generations · Food values · Food axiologies

10.1 Introduction

That the function of food does not consist merely in its being eaten and savoured is a known fact that has been codified by numerous disciplines. Indeed, a number of epistemological systems—frameworks of values and practices which produce attitudes, identity dynamics, fashion trends, and are continually challenging and being

B. Surace (✉)
Department of Philosophy and Educational Sciences, University of Turin, Turin, Italy
e-mail: b.surace@unito.it

S. Stano and A. Bentley (eds.), *Food for Thought*, Numanities - Arts and Humanities in Progress 19, https://doi.org/10.1007/978-3-030-81115-0_10

challenged through desemantisation, resemantisation, and processes of translation—rotate around the foodsphere.[1] In other words, today's "liquid society" (Bauman 1999) nurtures itself first on constant semiotisations of all that which concern food, and only subsequently on the materiality of food itself.[2] The perspectives and axiologies (or value assets) of these semantisations favour at times the nutritive aspects of food, and at other times their ethical or aesthetic ones. To recall Jean-Marie Floch's famous semiotic square of advertising valorisations (1990), the contemporary discourse on food constitutes a frenetic interchange between *base* and *use* values,[3] a "dribbling" which sometimes emphasises the practical-ludic relevance of a certain food or diet and the narcotisation of its ludic-utopian significance, and sometimes its exact opposite.[4]

It goes without saying that contemporary society globalised through new media intensifies in schizophrenic terms the excessive attention now being paid to food,[5] not infrequently generating unprecedented acrimony starting from the polarisation of food and eating behaviours and the solidification of identification practices connected to them.[6] These are processes that are now at their highpoint, but which are in reality founded on a consolidated relationship between food and media, and which are currently witnessing the affirmation of a new player: the younger generations.

10.2 Food, Adolescents, Media, Culture

The relationship between food and media does not, in fact, belong exclusively to modernity. Over and beyond prevalently scientific or descriptive contributions, there is a vast Greek and Latin literature that demonstrates how mealtime was far more than merely an occasion during which to nourish oneself, but a convivial event imbued with cultural connotations.[7] Immediate proof can be found by glancing at Trimalcione,

[1] The expression seems to be accepted today in the context of semiotic disciplines, and is a derivative of the broader concept of semiosphere (Lotman 1984). Cf. (Stano 2018a: 167).

[2] It is possible to distinguish between two plans of action in human experience. The first, characterized by what Cirese (1989) calls "fabrility", has to do with everything we do for our physical subsistence. The second, that of "signicity", has to do with our symbolic experience of the world. The thesis is therefore that food, even though it is something we need to survive, is actually approached in our experience above all in terms of signs, semiotically.

[3] The first is characterized as being a profound value, powerful and universal, capable of justifying the start of an engine of the narration and the action of the subject. In front of this value, the use value is characterized as having an instrumental and limited character" (García 2017: 123).

[4] The Floch model, created and developed to analyse advertising texts and function as a marketing tool, is actually useful for much wider purposes, regarding our way of enhancing the world starting from the macro-distinction between base and use values.

[5] In Türcke (2002) emphasis is placed on the fundamental role of "sensation" in modern societies, from a perspective which we consider akin to that which we posit here.

[6] In Surace (2018), for example, there is a reflection on the contrastive polarisation of the speeches of vegans and omnivores in the Italian mediasphere.

[7] Cf., for example, Wilkins (2000) or, for a more general approach, Wilkins and Nadeau 2015.

a character in Petronius' *Satyricon*. The episode in which he is the protagonist is so significant as to have codified the archetype of the *cena trimalchionis*.[8] This is a lucullan—i.e. sumptuous and extravagant—meal, whose length and extension in terms of quantity and quality of the (often exotic) food allow for the articulation of tales, incidents and reflections, some of which are also solemn in tone, in contrast to what the hedonism of the situation represented might lead one to suppose. It is ultimately a symbolic banquet, the ultimate Latin emblem of which is perhaps the Bacchanal, an orgiastic ceremony in which sex and food coalesce in a semiotic whole, where those partaking in the meal are called upon to share not only food but also experiences of life and entire systems of value. The same considerations may be applied to Plato's *Symposium*, also known as *Convivium*, where the frame of the banquet, offered by the tragic poet Agatone, becomes the occasion for the delineation of oratorical agony. What is more, the meal as an occasion for symbolic sharing is also a legacy of many religions, for example, in Christianity, where the moment of "breaking bread" and "sharing wine" constitutes a solemn declaration of human communion with transcendental reverberations.[9]

Food, as a catalyser of entire cultures, defines the highest common denominator on which still today widespread practices are founded. Tourism, for instance, is being increasingly influenced by considerations of food and cuisine. One of the most frequent questions asked of travellers on their return from a trip to foreign lands regards the food they tasted there. Expressions like "ethnic cuisine", "global cuisine", and "gastronomic tourism" are very popular, which can lead to the propagation of certain stereotypes. It is not uncommon, for example, on return from trips to China, and in perfect adherence to what Said calls "orientalism" (1978), to be besieged by questions about grasshoppers and scorpions, stemming from the conviction that the Far East is a land of avid insect-eaters.[10]

On the other hand, such experiences can lay a foundation for prolific cultural exchange. The allure of food can function as a kind of bridge or, in the terms used by Michel Foucault, a "heterotopy" (1967). The French philosopher defines the latter as the impossible junction of two otherwise unapproachable worlds. This is, in fact, what occurs every time we seek to approach geographically distant places or unknown cultures through what Simona Stano calls "the food of the Other" (2015), a food which is destined to nurture the mind even before the mouth (Ibid.). As Stano indeed maintains, reconnecting to Montanari and Flandrin (1997):

> Cooking is not a random assemblage of elements, but a unified and coherent system, there is a substantial difficulty in accepting, and sometimes even understanding the Other. Hence there is the need to "filter" what is unknown through one's own system of values, thus frequently distorting it, or at least adjusting it, reducing it to one's own criteria. (Stano 2015: 13)

In the light of such a complex context, it must be noted how recent is the food discourse associated with younger generations, including the so-called Generation

---

[8] A very impactful representation of this macro-episode can be found in the movie *Satyricon* by Federico Fellini (1969).

[9] Cf. Fieldhouse (2017).

[10] Cf. Kumar Dixit (2019).

Z.[11] Adolescence, a "zone of proximal development" (Vygotsky 1967)[12] between childhood and adulthood that first arose in the Romantic Age, is a stage at which strong identity construction takes place. Young people's fascination with food has led to a plethora of food activity—the production of texts as well as preparation of dishes—that has proliferated to the point of generating a *sub-genre*. Not only are these aesthetic texts about young people, but they often define them as well, by virtue of their tastes which, as Bourdieu (1979) teaches us, are responsible for the creation of social distinctions.

This involves a sort of overturning of the vision of childhood, naturally preluded here and there in earlier epochs, but which occurs to a great extent first with the birth of late Romantic literature for boys and girls, and then with the spread of an imaginary in which they are accorded an autonomy of thought that is no longer seen exclusively as something to be regimented within educational confines, but as a potential reservoir of creativity, worthy of note and not necessarily to be assessed in terms of deviance or adherence to a particular behaviour. Among the works that heralded this reversal can be cited, for example, the little-known poem *Contro la minestra* (*Against soup*), by eleven-year-old Giacomo Leopardi (1798–1837), one of Italy's most famous poets. Here the *enfant prodige* puts into verse his hatred for soup, channelling this into a poetic flow that inevitably prompts the adult world to ask itself whether it is right to nourish children with food they do not appreciate. It is by all effects a poem written by a child, with the eyes of a child, however erudite Leopardi may be, which exalts the perspective of the child, for whom the wholesomeness of the food has no value, as opposed to other aspects such as taste or texture:

> They say that when you [the soup] are good you can raise the dead;
> But this saying is worthy of people who are not very wise!
> Must one therefore be dead in order then to enjoy
> These benefits, which they say you possess?[13]

As mentioned above, it is roughly from the second half of the nineteenth century on that the juvenile perspective becomes more widespread thanks to literature for children and adolescents, for instance, novels such as *Alice in Wonderland* (Lewis Carroll 1865), *The Wonderful Wizard of Oz* (L. Frank Baum 1900), *The Jungle Book* (Rudyard Kipling 1894), and *The Little White Bird* (J. M. Barrie 1902, where Peter Pan appears for the first time).

With the subsequent advent of cinema there emerged an immediate relationship between film and food. An anthropological approach to early narrative Italian cinema reveals the material culture of food of the period (Cusano 2013), establishing a comparison between the ritual of cinema-viewing and that of eating, through the

[11] Generation Z seems to place the food issue as a priority in many respects. For example, about sustainability, cf. Kamenidou et al. (2019).

[12] Cf. also Steinberg et al. (2009).

[13] Verses 11–14, my translation of the original text: "Si dice, che risusciti, quando sei buona i morti; / Ma oh detto degno d'uomini invero poco accorti! / Or dunque esser bisogna morti per goder poi / Di questi beneficj, che sol si dicon tuoi?".

theories primarily of Lévi-Strauss (1958) and Douglas (1975), for whom food structurally can be seen as a language. From this way of reading early films, a route can be traced to contemporary cinema, passing through the cinematographic neo-genre of the 1980s, the *Food Movie*,[14] which is in fact significantly related, by means of narrative forms and "birth cohorts", to cinema about and for children *à la* Spielberg.

The late-twentieth century are years of great cinematographic ferment, with the making of action films and science fiction, fantastic adventures, romantic films and dramatic ones, all of which share a systematic focus on children and adolescents: *Star Wars* (George Lucas 1977), *Lampoon's Animal House* (John Landis 1978), *La Boum* (Claude Pinoteau 1980), *Friday the 13*th (Sean S. Cunningham 1980), *E.T. the Extra-Terrestrial* (Steven Spielberg 1982), *Flashdance* (Adrian Lyne 1983), *Wargames* (John Badham 1983), *A Nightmare on Elm Street* (Wes Craven 1984), *Sixteen Candles* (John Hughes 1984), *Neverending Story* (Wolfgang Petersen 1984), *Gremlins* (Joe Dante 1984), *The Goonies* (Steven Spielberg 1985), *Back to the Future* (Robert Zemeckis 1985), *Explorers* (Joe Dante 1985), *St. Elmo's Fire* (Joel Schumacher 1985), *Stand by Me* (Rob Reiner 1986), *National Dead Poets Society* (Peter Weir 1989), are examples of a tendency to mythicize the life of the adolescent,[15] as well as to give a new prominence to certain alimentary rituals, including parties, binges, picnics, the exchange of food as a pact of friendship, and snacks versus healthy meals.

10.3 New Generations, Values and New Media: A Model

With the ensuing advent of social media and platforms of video streaming, starting from YouTube and Twitch, younger generations have appropriated the imaginary of food and contributed to its modification. This is in part due to the history of philology outlined above, but also because of the diffusion of user-generated contents and prosumer logic, as well as the spread of new cultural markets and the massification of food as an aesthetic theme. These developments originated with certain types of television and cinema, but are also part of the new passion for documentaries and docufiction propagated by such services as Netflix, where food-focused series, reality shows and cooking competitions are the order of the day.

In this cross-medial and cross-textual environment, the memetic dynamics determine the dominance of the meta-communicative logic of virality whereby often the formats are repeated identically or with minimal variations on theme.[16] Here the YouTuber is essentially a young individual producing content for other young people or even for children. Among the numerous genres which populate the platform

[14] On food and cinema see Bower (2004), Zimmerman and Weiss (2010). More in general on food and popular culture cf. Lebesco and Naccarato (2018).

[15] Cf. Shary (2002).

[16] The virality metaphor is introduced in the human sciences starting from Dawkins' memetic theories (1976), which are rather controversial today because of their reductionist approach, but have paved the way for a long series of studies. A significant compendium in terms of methodologies and approaches can be found in in Marino and Thibault (2016).

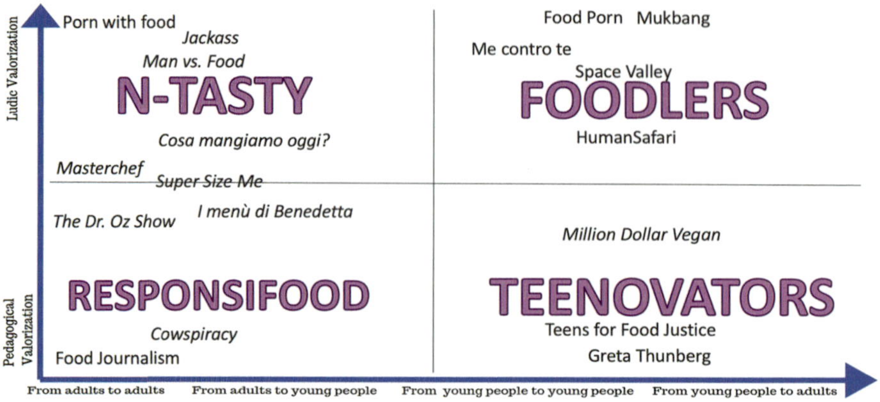

Fig. 10.1 Graph of food valorisations in the media

are those related to food, including playing with food, "testing-tasting" food that is considered strange, reviewing and therefore judging food, justifying certain diets and demolishing others. Media broadcasters subsequently appropriate the particularly successful formats and respond by proposing spin-offs.

It is possible to map these contents by arranging them along certain axes (Fig. 10.1):

(1) Anagraphic axis of orientation of the meaning of food, which articulates the following dimensions: (i) a media product conceived/realised by adults for adults; (ii) a media product conceived/realised by adults for young people; (iii) a media product conceived/realised by young people for young people; (iv) a media product conceived/realised by young people for adults.
(2) Axis of dominant values: (i). a pedagogical dominant, that is, the production of texts which convey a critical-practical valorisation of food in various terms: nutritional, environmental, cultural, economic, and so on; (ii) a ludic dominant, that is, the production of texts that convey a utopian-recreational valorisation of food, productions where food is playful, fun, strange, absurd, exotic, and so on.

Moving along the horizontal axis from the vertex, we find the media products realised by adults for adults. At the bottom, the pedagogical valorisation is conspicuous, which includes serious documentaries or broadcasts that aim to convey informative content of some kind. As we rise, the ludic components become gradually more present. The greater the distance, the more these recreational components overwhelm the pedagogical ones. Similarly, by moving from left to right we move to different targets and broadcasters.

This kind of schematisation immediately emphasizes the fundamental dynamic nature of the framework; on one hand necessarily constituting a boundary, while on the other being transitory. A documentary like *Cowspiracy: The Sustainability Secret* (Kip Andersen and Keegan Kuhn 2014), for example, even if fixed on the broadcaster's side (conceived and realised by adults), is today viewable by adults and by young people alike. Similarly, the culinary *vlogs* of young YouTubers, realised by young people and designed primarily for young people, provide entertainment for adult audiences too.

Bearing in mind these premises, it is possible to isolate a couple of cases for each quadrant of this model, and rename the quadrants themselves as follows:

(1) N-TASTY: The adult sphere on the productive side, adult-youth interpolation on the receiving end, ludic valorisation. Here we can find television shows and films in which food is a primarily aesthetic element or is symbolised on certain axiologies. The radicalisation of this quadrant, for instance, is real "food pornography", where food is valorised on the basis of its aphrodisiac or plastic components and the Model Spectator is clearly an adult individual, who adores watching little games with cream or similar haptic-gustative performances; moving tensively—with flexibility—and therefore elasticizing both the Model Spectator and the valorisation, further texts emerge. If, for instance, we maintain the ludic component unaltered at its peak, but move towards products that are gradually targeted by both adults and youths, then we can include television shows like *Man versus Food,* where food becomes an instance of exorbitance and excess, and an element of challenge. This type of format is, in a typical movement of reciprocal contamination, inspired by the diffusion in the YouTube-sphere of the aesthetics of culinary challenges, but also constitutes in its turn the serial consolidation of a certain type of script. If, instead, we move from the pole of ludic valorisation towards that of pedagogical valorisation, then we see the appearance of shows like *Masterchef*. On one hand, this is completely designed to present the spectator, whether adult or young, with a series of challenges, where the purpose of food is purely to entertain. On the other hand, food also becomes a vehicle for the conveyance of moral values such as the importance of hard training, seasonality, creativity, and in the latest issues such as eco-sustainability as well, this last rendered with profilmic emphasis on the doggy bag with which the competitors are provided in order not to waste the food left over from the trials. Similar considerations can be moved towards *Cosa mangiamo oggi?* (*What are we eating today*?), a YouTube and Twitch channel where the two protagonists propose different approaches to the foodsphere, playfully and light-heartedly but at the same time treating food as something which not simply comprises nutrition, but has to do with much broader terrain: Food can be of different qualities, has a history, has traditional preparation processes to be followed, and is a complex cultural object, to be known and respected. This channel can thus act as a junction with the next quadrant (Fig. 10.2).

(2) RESPONSIFOOD: The adult sphere on the productive end, adult-youth interpolation on the receiving end, pedagogical valorisation. Descending further, one crosses into the second quadrant, where the ludic component gradually decreases and is replaced by programmes and shows in which, albeit in tones that are not exactly serious, food is increasingly valorised in more exclusively

Fig. 10.2 "Cosa mangiamo oggi?" (Italian YouTube and Twitch Channel)

pedagogical terms. It is therefore no longer an element of play or the core of an aesthetics, but something that can be either nourishing or harmful, or that considers the environment, specific production systems and so on. In this case, too, the radicalisation is in relation to texts where food is valorised without any aesthetic component and treated purely as nutrition. Within this quadrant a certain flexibility ensures that even when food is treated seriously and with the purpose of informing or conveying specific critical content, this may nevertheless be thematised, ludicised or ideologised, for instance, in documentaries such as *Super Size Me* (Morgan Spurlock 2004) or in television programmes such as Italian *Benedetta's Menus,* where the presenter at times cooks together with her daughters, transforming the act of cooking into a game, though with the scope of producing dishes that are simultaneously fun and healthy. Moving to contents created by young broadcasters, we can see how here, too, there is an abundance of texts and an intricate system of variations on the theme. This is due not only to the increasingly relevant role accorded to the younger generations in the post-modern age, but also to their possibility through the new media of autonomously producing their own content and disseminating it, from a bottom-up logic, without necessarily having much capital at their disposal. The fact that food itself—through a paradigm of continuous exchange of fashions and trends, often generated also by certain algorithms like those of YouTube which privilege different contents at different times—is one of the most popular elements in the context of new media, reveals the enormous importance of the alimentary sphere for Generation Z, and even before them for the so-called "Millennials". Both these contiguous generations are, in fact, denoted by an unprecedented sense of responsibility and by a widespread critical attitude, and also an awareness that technology has fostered new forms of power. The result is a society that is increasingly inhabited by reviewers, automatic evaluation systems, communities dedicated to one or other particular theme, opinion leaders selected by means of online channels and platforms, new mechanisms of trust, loyalty and approval. Traditional pyramidal broadcasting systems are disappearing in favour of new forms of narrowcasting narration, just as parental and school authority is being partially reconfigured in the face of competition from an ever-denser semiotic system from which to draw one's knowledge and one's encyclopaedias of reference. The new generations are developing a critical sensibility that today focuses on themes of food and what gravitates around it.

(3) FOODLERS: The young sphere on the productive end, youth-adult interpolation on the receiving end, ludic valorisation. In this case, in the upper left-hand corner we have the highest ludic valorisation and the almost exclusive targeting of young people. The elastic modulation of the ludic component entails a shift in axiologies: food as matter whose edibility is not necessarily the ultimate value; as plastic matter, eminently visual and haptic. It is possible to linger, for example, on the case of *Me Contro Te (Me against you)*, Italian YouTubers whose videos are clearly addressed at an audience of children. In their video catalogue, there are titles such as "Let's try American snacks!", "Real food versus Pop-in Cooking", "Let's mix all the cereals together", "Let's buy everything on the McDonald's menu! Sandwiches you've never seen before". From their very titles it is immediately visible how food is treated here as a spectacular element of pure entertainment, which can be experienced playfully, especially when it is taken to a level of consumer excess. Exaggerated combinations, like the union of various types of cereals, exoticism, as in trying snacks from distant lands, the display of obviously excessive consumption, as in the case of ordering an entire menu, are all patterns typical of this type of ludic valorisation, which is also expressed by the formal rhetoric of the videos (and often motivated by the dynamics of sponsorship and monetisation of the videos). The videos are loud, frantic, filled with visual and sound effects that have been added post-production. There is an insistent use of the *shaky cam* and, in general, totally codified, normal operations like ordering food from home are presented as great fun and highly unusual. The food, in this case where the *ennui* of everyday life is disguised as boundless pleasure, is fully embedded in show-business apparatus, and also in formal terms is arranged and filmed with close-ups or total fields which stage scenes of abundance, typical of that which is today called *foodporn*,[17] an aesthetics of alimentary excess that is in vogue on Instagram, YouTube and many other social media.
Remaining at this level of ludicity, but moving towards a more hybrid target containing both youngsters and adults, we can mention the crew of *Space Valley*, Italian YouTubers who literally play with food, unhinging it from a certain system of rules and re-contextualising it as matter that can be treated in

[17] Cf. Stano (2018b).

Fig. 10.3 Space valley (Italian YouTube Channel)

numerous ways. Their videos contain a succession of unusual food combinations, verging on disgusting (salmon with cocoa cream), uncommon cooking techniques (egg cooked on an iron) and so on. The *Space Valley* crew, by means of sophisticated screen work, direction and editing, stage culinary challenges where food is consumed above all with one's eyes, and then vicariously through the protagonists who entertain the audience by tickling a certain masochistic propensity, as happened in the television show *Jackass* whose protagonists practised, among other things, coprophagia[18] (Fig. 10.3).

Similarly, there is the *mukbang,* initially popular in South Korea but spreading to the rest of the world, the final frontier of food porn where a web star of tender years eats dangerously huge quantities of food, in a sort of competitive logic, livestreamed in front of an audience. It is a spectacle of food excess and deturpation, in which the spectators' voyeurism, in a "sadistic ceremony" (Raciti 2019), is exalted by a sensation of control. The ludic dimension is so radicalised as to degenerate into something that is almost punitive. The food that is bolted down, displayed to effect by the camera first appears as an appealing cornucopia, but begins to produce viewer discomfort, becoming an instrument of torture as the act of binge eating cognitively transforms it into a toxic stew. The para-pornographic nature of this textual genre is exacerbated by its being combined with formal filming specificities, such as high resolution details of the food, and by a particular dialectic of depth of field which places the eater in the background and the food in the foreground. The genre also features tight close ups of the face, the only part of the subject which is usually shown, which expresses increasing agony as the video proceeds, and by other specificities such as the so-called ASMR (Autonomous Sensory Meridian Response), generated by means of special microphones which capture and amplify the

[18] Cf. Surace (2019).

sounds of chewing and swallowing, paradoxically found extremely relaxing by some and highly irritating by others (the so-called "misophonists").[19] This radicalisation clearly entails a total de-responsabilisation in the face of food and of one's own body, a contrast to an opposite radicalisation that may lead to psycho-pathological phenomena like orthorexia nervosa,[20] wherein the attempt to eat in the best possible way, respecting all the (medical, nutritional, environmental, economic, cultural, etc.) rules and a specific set of values, produces obsessive manias.

Over and beyond this type of video, the two mentioned previously are significant because while food is a protagonist it is not, however, totalising. Both *Me Contro Te* and *Space Valley* are YouTubers who do not make videos dedicated exclusively to food. It is quite a constant presence, but organically inserted in a broader context where the YouTubers, with their dynamism and overall compelling qualities, are the ultimate protagonists. The same happens for travel vlogging channels in which the culinary element often appears, as in the case of *HumanSafari,* where the vlogger roams the world with his videocamera, giving prominence to what he eats as he does to the places he visits.

(4) TEENOVATORS: The young sphere on the productive end, youth-adult interpolation on the receiving end, pedagogical valorisation. This is the last quadrant, which features pedagogical valorisations proposed by the new generations themselves. Here food is valorised from a critical-practical point of view and, at least today, topics related to environmental crises such as climate change are dominant. It is immediately apparent that the target is unlikely to be composed exclusively either of youngsters or adults. The messages produced by these types of texts, in fact, are directed at both anagraphic regimes simultaneously. If Greta Thunberg is now a worldwide symbol of the movement which gives out a certain sort of message, this already explains how the alimentary dimension is actually inserted in a broader context, which is not connected to food. It is a vision of a system, which over and beyond Thunberg finds space, for example, in the YouTube channel *Million Dollar Vegan,* an organisation which promises to donate one million dollars to charity if Pope Francesco promote a vegan Lent.[21] Although the organisation is clearly vast and managed by adults, the promoter whose face is the symbol for the communication of this project is Genesis Butler, a twelve-year-old environmental activist, the youngest person ever to have taken the *Tedx* stage.

The vegan diet is therefore here presented as a radical choice, disengaged from a vision of food as pleasure and rooted in an ethical vision of food as responsibility, and the choice of a young face is useful in conveying a certain type of rhetoric. Butler travels and tells stories through her own eyes, and these stories are narrated with particular attention to form. In the video "Meet the Italian farmer who stopped killing and started saving animals", for example,

[19] On the ASMR phenomenon cf. Poerio (2016).

[20] On orthorexia nervosa cf. Brytek-Matera et al. (2016).

[21] https://www.milliondollarvegan.com/pope-francis-responds-to-12-year-old-activist/.

Butler's interview with an ex-breeder, now vegan, is interspersed with close-ups of baby animals, the smiling, happy face of the young activist in their midst, all accompanied by jovial background music. Genesis Butler is obviously not the only one inserted in this quadrant, which has been in fact re-named to that of the *teenovators* on account of the *Seed and Chips* initiative in 2018, when the new generations, through a series of alimentary themes, "sit around a table" in order to propose their idea of innovation in connection with food at the Global Food Innovation Summit in Milan. These types of operations are currently the order of the day, as is shown by other instances like *Teens for Food Justice* and so on.

These quadrants are not themselves static, but are characterised by flexibility—that is, they contain a "tensivity" in which are articulated mediatised products that are more polarised towards an adult dimension, and products that instead trend more towards the adolescent sphere; the same applies to the axis of valorisation. Moving towards the centre, a situation of "generalist" balance is reached.

This brief model, which is to be understood as an initial attempt to create some order in the complex contemporary mediated foodsphere, shows how the insertion of the collective actor constituted by the new generations into the global alimentary debate has entailed a significant redefinition of the axiologies connected to food.

References

Barrie, J.M. 1902. *The Little White Bird*. London: Hodder & Stoughton.

Baum, L. Frank. 1900. *The Wonderful Wizard of Oz*. George M. Hill Company: Chicago.

Bauman, Zygmunt. 1999. *Liquid Modernity*. Cambridge: Polity.

Bourdieu, Pierre. 1979. *La Distinction. Critique Sociale du Jugement*. Paris: Les éditions de minuit.

Bower, Anne L. 2004. *Reel Food. Essays on Food and Film*. New York-London: Routledge.

Brytek-Matera, Anna, Lorenzo M. Donini, Magdalena Krupa, Eleonora Poggiogalle, and Phillipa Hay. 2016. Orthorexia nervosa and self-attitudinal aspects of body image in female and male university students. *Journal of Eating Disorders* 4 (16). https://doi.org/10.1186/s40337-015-0038-2.

Carroll, Lewis. 1865. *Alice in Wonderland*. London: MacMillan.

Cirese, Alberto M. 1989. *Segnicità, Fabrilità, Procreazione. Appunti Etnoantropologici*. Rome: CISU.

Cusano, Jessica. 2013. Fame di film. Cibo e generi nel cinema italiano. *Immagine. Note Di Storia Del Cinema* 8: 114–141.

Dawkins, Richard. 1976. *The Selfish Gene*. Oxford: Oxford University Press.

Douglas, Mary. 1975. *Implicit Meanings. Essays in Anthropology*. London: Routgledge & Kegan.

Fieldhouse, Paul. 2017. *Food, Feasts, and Faith. An Encyclopedia of Food Culture in World Religions*. Santa Barbara-Denver: ABC-CLIO.

Floch, Jean-Marie. 1990. *Semiotique, Marketing et Communication. Sous les signes, les strategies*. Paris: Presses Universitaires de France.

Foucault, Michel. 1967. Des Espace Autres (Conférence au Cercle D'études Architecturales, 14 Marzo 1967), *Architecture, Mouvement, Continuité* 5 (Octobre 1984), 46–49.

García, Lidia P. 2017. Identity: Linguistic-Semiotic Analysis. In *Human Development III*, eds. Francisco M. Moreno Lucas, Maria I. Rojas Marin, and Marco Bracci, 117–134. Cambridge: Cambridge Scholars Publishing.
Kamenidou, Irene C., Spyridon A. Mamalis, Stavros Pavlidis, and Evangelia-Zoi G. Bara. 2019. Segmenting the generation Z cohort university students based on sustainable food consumption behavior: a preliminary study. *Sustainability* 11: 837.
Kipling, Ruyard. 1894. *The Jungle Book*. London: Macmillan.
Kumar Dixit, Saurabh (ed.). 2019. *The Routledge Handbook of Gastronomic Tourism*. London-New York: Routledge.
LeBesco, Kathleen, and Peter Naccarato (eds.). 2018. *The Bloomsbury Handbook of Food and Popular Culture*. London-New York: Bloomsbury.
Lévi-Strauss, Claude. 1958. *Anthropologie Structurale*. Paris: Plon.
Lotman, Juri M. 1984. O semiosfere. *Sign System Studies* 17: 5–23.
Marino, Gabriele, and Mattia Thibault (eds.). 2016. *Virality, Monographic Issue of Lexia—Journal of Semiotics*, 25–26. Rome: Aracne.
Montanari, Massimo, and Jean-Louis Flandrin, eds. 1997. *Storia Dell'alimentazione*. Rome-Bari: Laterza.
Poerio, Giulia. 2016. Could insomnia be relieved with a youtube video? The relaxation and calm of ASMR. In *The Restless Compendium. Interdisciplinary Investigations of Rest and Its Opposites*, eds. Felicity Callard, Kimberley Staines and James Wilkes, 119–128. Cham: Palgrave.
Raciti, Giulia. 2019. La sadica cerimonia del cibo nel Mukbang. *E|C. Rivista dell'Associazione Italiana di Studi Semiotici*. http://www.ec-aiss.it/includes/tng/pub/tNG_download4.php?KT_download1=89ab53b6793403a436e6a487144d054b. Accessed 29 Apr 2020.
Said, Edward W. 1978. *Orientalism*. New York: Vintage Books.
Shary, Timothy. 2002. *Generation Multiples. The Image of Youth in American Cinema since 1980*. Austin: University of Texas Press.
Stano, Simona. 2015. *Eating the Other. Translations of the Culinary Code*. Newcastle upon Tyne: Cambridge Scholars Publishing.
Stano, Simona. 2018a. *I Sensi Del Cibo. Elementi di Semiotica Dell'alimentazione*. Rome: Aracne.
Stano, Simona. 2018b. *Mauvais à regarder, bon à penser*: il *food porn* tra gusti e disgusti. *E|C. Rivista dell'Associazione Italiana di Studi Semiotici*. http://www.ec-aiss.it/includes/tng/pub/tNG_download4.php?KT_download1=9abbb72a5350908ea27a5a42555e334a. Accessed 29 Apr 2020.
Steinberg, Laurence, Deborah Lowe Vandell, and Marc H. Bornstein. 2009. *Development: Infancy Through Adolescence*. Belmont: Wadsworth.
Surace, Bruno. 2018. Pokémon and the Peta. Viral Extremeness as a Semiotic Strategy. In *Virality and Morphogenesis of Right Wing Internet Populism*, ed. Eva Kimminich, Julius Erdmann, and Amir Dizdarevič, 151–162. Frankfurt: Peter Lang.
Surace, Bruno. 2019. Elementi di coprosemiotica. *E|C. Rivista dell'Associazione Italiana di Studi Semiotici*. http://www.ec-aiss.it/includes/tng/pub/tNG_download4.php?KT_download1=26ff4523c71fa0c2aa88251593f3a5fe. Accessed 29 Apr 2020.
Türcke, Christoph. 2002. *Erregte Gesellschaft. Philosophie der Sensation*. Munich: C. H. Beck.
Vygotsky, Lev S. 1967. Play and its role in the mental development of the child. *Soviet Psychology* 5 (3): 6–18.
Wilkins, John. 2000. *The Boastul Chef. The Discourse of Food in Ancient Greek Comedy*. Oxford: Oxford University Press.
Wilkins John, and Robin Nadeau, (eds.). 2015. *A Companion to Food in the Ancient World*. Chichester: Wiley Blackwell.
Zimmerman, Steve, and Ken Weiss. 2010. *Food in the Movies*. Jefferson and London: McFarland & Co.

Bruno Surace is a Ph.D. in Semiotics and Media at the University of Turin, Post-Doc Research Fellow in the ERC Project FACETS (Face Aesthetics in Contemporary E-Technological Societies), Adjunct Professor of General Semiotics and of Cinema and Audiovisual Communication at the University of Turin and Lecturer at the Renato Einaudi Interuniversity College. He is the author of the monograph *Il destino impresso. For a theory of destinality in cinema* (Turin, Kaplan 2019), and co-curator of the volumes *I discorsi della fine. Catastrofi, disastri, Apocalissi* (Aracne 2018) and *The Waterfall and the Fountain. Comparative Semiotic Essays on Contemporary Arts in China* (Aracne 2019). He has published articles in several peer reviewed journals, as well as book chapters. He has participated as a speaker at numerous conferences in Europe, China and the United States.

Part III
Nutrition and Culture

Chapter 11
Beyond Nutrition: Meanings, Narratives, Myths

Simona Stano

Abstract As Barthes (Annales ESC, XVI(5): 977–986, 1961) effectively pointed out, food is "not only a collection of products that can be used for statistical or nutritional studies. It is also, and at the same time, a system of communication, a body of images, a protocol of usages, situations, and behaviors" (ET 1997: 21). This has become even more evident in present-day "gastromania": not only do we eat food, but also and above all we talk about it, we comment on it, and we share its images on various social networks, thus investing it with multiple meanings and values that in turn mediate our gastronomic experiences. This phenomenon has become progressively more expansive, encompassing the sphere of nutrition. Going beyond the purely dietetic and medical domains, the link between food and health has become an unavoidable element of TV programmes, newspapers, magazines, social networks, advertising, marketing, and other forms of communication. Thus a series of food "myths" have proliferated, with evident impact on consumers' choices and behaviours. What is more, the role played by media companies, marketing operators and various other public and private actors in the negotiation of food meanings and practices has further increased, pointing to the need for deeper consideration of the processes of signification and valorisation brought about by the discursive strategies adopted for communicating food in the political, journalistic, regulatory and even scientific domain. This essay investigates such dynamics by considering relevant literature in the related fields of research and analysing some interesting case studies.

Keywords Nutrition · Nutritionism · Myth · Value · Meaning

This project has received funding from the European Union's Horizon 2020 research and innovation programme under the Marie Sklodowska-Curie grant agreement No 795025. It reflects only the author's view and the European Research Executive Agency is not responsible for any use that may be made of the information it contains.

S. Stano (✉)
Department of Philosophy and Educational Sciences, University of Turin, Turin, Italy
e-mail: simona.stano@unito.it

S. Stano and A. Bentley (eds.), *Food for Thought*, Numanities - Arts and Humanities in Progress 19, https://doi.org/10.1007/978-3-030-81115-0_11

11.1 Introduction

As its title suggests, this essay presents a thought-provoking reflection on the relation between nutrition, which is one of the aspects of food mostly associated with the material level, on the one hand, and culture and meaning, on the other hand. To this purpose, it deals with some crucial issues related to nutrition and its understanding in contemporary foodscapes, mainly adopting a semiotic approach.

Let us start, as usual in semiotics, by focusing on the very definition of our object of research: *food*. The Collins Dictionary defines it as "any nourishing substance eaten or drunk to sustain life, provide energy, and promote growth" (2020), thus highlighting the connection of eating to other basic needs, namely the sustenance of our body and its vital functions. Merriam-Webster further enhances such a conception, also stressing the chemical and nutritional composition of food: "Material consisting essentially of protein, carbohydrate, and fat used in the body of an organism to sustain growth, repair, and vital processes and to furnish energy" (2020).

From a semiotic point of view, it is interesting to consider such definitions in light of the so-called "square of valorisations" introduced by Floch[1] (1990): the idea of food highlighted by the above-mentioned definitions is evidently linked to a *practical valorisation*, since eating is inserted in a second-level narrative programme aimed at achieving other values such as survival, growth, and energy. This is precisely what we generally call *nutrition*, that is to say, "the process of providing or obtaining the food necessary for health and growth" (Merriam-Webster 2020), with dietetics being "the science or art of applying the principles of nutrition to feeding" (Merriam-Webster 2020).

However, as Roland Barthes effectively showed, food is "not only a collection of products that can be used for statistical or nutritional studies. It is also, and at the same time, a system of communication, a body of images, a protocol of usages, situations, and behaviors" (1961; Engl. Trans. 1997: 21). In fact, not only do we eat foods that allow us to maintain or recover health, but, more or less consciously, we eat foods that express our identity (in accordance with the famous idea "bon à penser, bon à manger" described by Lévi-Strauss (1962)), in a *utopian valorisation*. Also, and not less importantly, eating can give us pleasure (in terms of flavours, but also of visual appraisal, ludic entertaining, etc.), thus inserting food in a *ludic* or *aesthetic*

---

[1] Projecting the categorical distinction between use values and basic values—described by Greimas and Courtés (1979) through the well-known example of the banana that the monkey tries to reach (basic value) and the stick he uses to execute such a process (use value)—onto a semiotic square, the French semiotician identifies four major "valorisations" of products in advertising: the practical valorisation puts the emphasis on utilitarian or use values, presenting its object as a tool, that is to say, as a means to reach another goal; the utopian valorisation rather stresses existential or basic values such as life, identity, freedom, etc.; the ludic or aesthetic valorisation is the negation of the practical one, and emphasises values such as gratuity and refinement; finally, the critic valorisation, which represents the negation of the utopian one, corresponds to a logic of calculations and interests (costs/benefits or quality/price ratios). Although developed in reference to advertising, this model in fact concerns all processes consisting in the projection of specific values onto specific objects or practices.

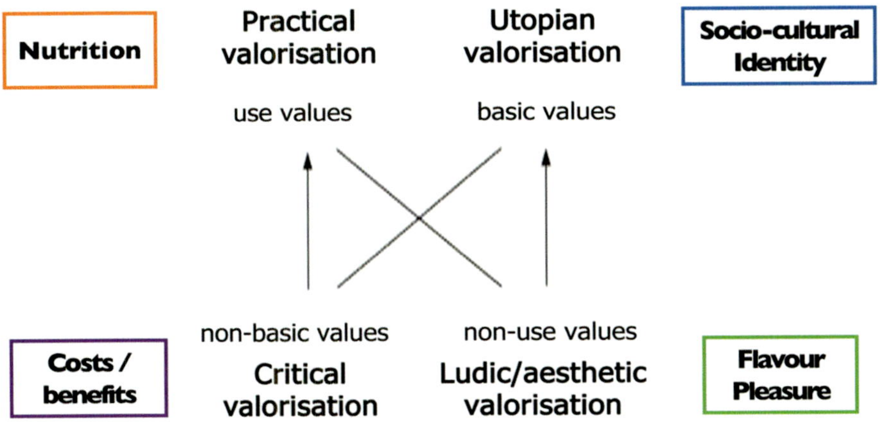

Fig. 11.1 The square of valorisations of food

valorisation. Sometimes, moreover, a costs/benefits ratio comes into play, resulting in a *critic valorisation* (Fig. 11.1).

It is important to note that such valorisations are not irreconcilable, but rather complementary, as the history of nutrition itself highlights. We will consider some interesting examples in the following paragraphs, adopting a diachronic approach that will lead us to highlight the role and conception of nutrition, and more generally of food, in contemporary foodscapes.

11.2 Food, Health, and "Myths"

Eating healthy food is not a novel concern. The link between food and health has always played a crucial role in human societies: because of their properties, various food substances and practices have historically been used to prevent and cure body alterations. What has changed is rather how various societies and cultures have understood the link between food and health, and the systems of values they have projected on such a link.

As Laudan (2004) interestingly points out, for instance, before the middle of the seventeenth century, cooking represented the basic metaphor for vital systems: the sun "cooked" seeds into plants and the latter into fruits and grains; humans then "cooked" such substances into edible dishes; and finally their internal organs "cooked" the ingested food by digesting it to produce energy and vital substances, excreting wastes and returning them to the soil, where the cosmic culinary cycle began again. This led to categorising food into easily digestible foods, which were preferred in healthy diets, and those substances difficult to digest, which had to be avoided as much as possible but could also be used as drugs, if well administered.

After 1650, the metaphor of cooking was replaced by that of "fermentation", which was associated with putrefaction, distillation and the interaction of acids and salts. Although always starting with seeds and ending with the release of waste, the cosmic culinary cycle of that era therefore involved a different approach to food itself: "the cosmos was still a kitchen but was now equipped with brewers' vats, and the human body held miniature copies of that equipment" (Ibid.: 15). As a consequence, a preference emerged for food products that fermented readily (e.g. oysters, anchovies, green vegetables, mushrooms and fruits) and thus did not need complicated preparation in the kitchen to be pre-digested, with evident changes in the "healthy diet" of that time. Different substances were thought to be healthier and hence adopted in people's diets based not merely on their material composition, but rather depending on the values attributed to such material aspects.

This fact has become even more evident with the recent inclusion of the Mediterranean Diet in the UNESCO Representative List of the Intangible Cultural Heritage of Humanity. Originated from the scientific field, in the wake of medical research (Keys and Keys 1975; Keys 1980) correlating the low incidence of cardiovascular diseases among the inhabitants of specific areas and a particular nutritional regime, the Mediterranean Diet was initially mainly defined by the use of certain ingredients and specific techniques of preparation of food. As time went by, the focus of attention moved from the simple definition of healthy rules regulating nutrition to the social and cultural implications of the particular "lifestyle" that came to be identified with the Mediterranean Diet (with special emphasis put on conviviality, sharing and tradition, see UNESCO 2012, 2013). It thus became a general model, a sort of universal form as indicated by the attempt to adapt other—and sometimes even very different—food codes to it (cf. Stano 2018).

While material changes are certainly important, the significance goes well beyond it. Rather, specific "myths", intended—in Barthes' (1957) terms—as "second order semiological systems" or "meta-languages" naturalising specific views and ideologies so that they remain unperceived, are at play. Food issues are at the crossroads of public institutions, media companies, marketing operators and various other public and private actors, which participate in a vibrant battle of negotiation for the attribution of meanings of specific products and practices. In contemporary foodscapes, such processes have become not only widespread, but also extremely important, as we will discuss in detail in the following paragraphs.

11.3 Nutrition(ism) in Contemporary Gastromania

In the last decades the so-called phenomenon of "gastromania" (Marrone 2014) has progressively intensified, expanding to encompass the sphere of nutrition: not only do we eat food, but also and above all we talk about it, we comment on it, we share its images on various social networks, and so forth, thus investing it with multiple meanings and values that in turn mediate our gastronomic experiences. Going beyond the purely dietetic and medical domains, the link between food and health has become an unavoidable element of TV programmes, newspapers, magazines, social networks, advertising, marketing, and other forms of communication. Thus nutrition has come to embody itself a cardinal principle of contemporary food myths, triggering interesting meaning-making processes.

Let us consider, for instance, the obligation to include information on food values, including energy density (calories) not only on food labels, but also in qualifying restaurants, in the United States[2] and other countries, as well as the increased attention paid to nutritional information by both the food industry and consumers. This is part of a more general trend, which has been described and critiqued under the name of "nutritionism":

> Nutrition scientists, dieticians, and public health authorities—the nutrition industry, for short—have implicitly or explicitly encouraged us to think about foods in terms of their nutrient composition, to make the connection between particular nutrients and bodily health, and to construct "nutritionally balanced" diets on this basis. ... I refer to this nutritionally reductive approach to food as the ideology or paradigm of nutritionism. This focus on nutrients has come to dominate, to undermine, and to replace other ways of engaging with food and of contextualizing the relationship between food and the body. Nutritionism is the dominant paradigm within nutrition science itself, and frames much professional—and government—endorsed dietary advice. But over the past couple of decades nutritionism has been co-opted by the food industry and has become a powerful means of marketing their products. (Scrinis 2008: 39)

The nutritional dimension, which in itself would not constitute a culinary text (Marrone 2016: 188), becomes in such a discourse a real system of signification of food—and, more specifically, a "system of classification" (cf. Douglas 1972) of edible and inedible substances, although—as Scrinis (2008, 2013) denounces—in a reductionist manner and on the basis of a process of de-contextualisation, simplification and exaggeration of the role of nutrients.

In her book *Food Politics: How the FoodIndustry Influences Nutritionand Health*, Nestle (2013 [2002]) reports an interesting example of food communication that seems highly significant in relation to such a phenomenon. In 1999, Heinz released an advertisement showing a bottle of ketchup with the headline "Lycopene may help reduce the risk of prostate and cervical cancer*", using the asterisk to refer to a review article on the health benefits of such a component and thus associating the product with cancer prevention. In addition to the lack of scientific substantiation of the reported claim, the evident problem with this message concerns precisely the isolation of a specific nutrient, which is de-contextualised and given meaning independently from the other components of the food it is part of:

> Ketchup contains processed tomatoes, sugars, and salt (in that order) and could hardly be considered a health food, not least because it typically is used as a garnish for hamburgers and fried potatoes. The advertisement singled out one component of ketchup, lycopene (a plant pigment naturally present in tomatoes and other fruits and vegetables), and it clearly associated ketchup with cancer prevention by including a prominent endorsement from the Cancer Research Foundation of America". (Nestle 2013 [2002]: 334)

---

[2] The reference is to the so-called "calorie count laws", requiring qualifying restaurants (i.e. with over 20 locations nationwide) to post food energy and nutritional information on the food served on menus and menu boards.

Despite these issues, such a communicative strategy proved quite effective, and the brand reported a 4% increase in market share as a result of the ad.

Evidently, it is not the food item that is meaningful in such a paradigm, but rather one or a few of its constituents, independently from their—symbolic, and also material—context. In fact, a number of scholars and movements have increasingly pointed out this fact. The Spanish movement "Realfooding",[3] for instance, denounces:

> Hablamos en términos de hidratos de carbono, grasas, proteínas, vitaminas, minerales... cuando todo eso no tiene nada que ver con la salud. La salud tiene que ver con los alimentos, que son mucho más que la suma de sus nutrientes y calorías. [*We speak in terms of carbohydrates, fats, proteins, vitamins, minerals... but all that is not related to health. Healthis related to food, which is much more than the sum of its nutrients and calories*, our translation.]

This has resulted in the evident increase of "free from" (gluten, lactose, etc.) products and diets, as well as in the large adoption of "functional foods" (i.e. foods containing health-giving additives and promising medicinal benefits) and dietary supplements, with evident impact on the axiologisation of food. The nutritional discourse has become a symbolic logic itself, a "mythology" regulating the food realm, mainly based on a negative logic of deprivation. While there have always been taboos[4] in the food realm, today prohibition itself has become a basic choice. Diets have thus ceased to be a measure of well-being to become a real condition of being. In other words, as Niola (2015) argues, if once we were the ones making our diet, now it's "our diet that makes us", neglecting that, in principle—as Marrone (2019) reminds us—dietetics should not eliminate but rather organise foodstuffs; it should not prohibit but rather regulate food choices; it should not restrict but rather direct one's possibilities.

Let us consider, for instance, the present-day "gluten-free mania". Until about a decade ago, little attention was paid to the coeliac disease—an autoimmune disorder properly identified only in the 1950s,[5] which occurs exclusively in genetically predisposed subjects—, and people rarely seemed to give gluten much thought. Driven by books like *Wheat Belly* by cardiologist William Davis, *Grain Brain: The Surprising Truth About Wheat, Carbs, and Sugar—Your Brain's Silent Killers* by neurologist David Perlmutter, and other texts strongly criticised at first, and later a number of posts and comments on social media, a gluten-free movement originated and largely spread across the globe. It also gave birth to a real syndrome, called "(non–coeliac) gluten sensitivity" (NCGS), which is not genetic and does not involve autoimmune comorbidities (see in particular Ludvigsson et al. 2013; Volta et al. 2013; Elli et al. 2015; Fasano et al. 2015; Schuppan et al. 2015; Vriezinga et al. 2015), and for these reasons is still matter of debate within the scientific community. Yet it represents an increasing factor in today's "orthorexic societies"[6] (cf. Nicolosi 2006/2007; 2007),

[3] https://realfooding.com.

[4] Whose importance and significance have been effectively illustrated by scholars such as Douglas (1966).

[5] In 1956, British physician Margot Shiner pioneered the use of intestinal biopsy capsules, leading to conclusive evidence that gluten affected the intestinal mucosa in celiac patients.

[6] Building on Bratman's definition of "orthorexia nervosa" (see in particular Bratman 1997 and Bratman and Knight 2001) as a psycho-cultural syndrome consisting in a state of hyper-attention

often based on self-diagnosis or unorthodox diagnostic practices. While the material dimension is certainly crucial to the coeliac disease, whereby gluten as such cannot be digested by celiac people, gluten sensitivity seems to be more linked to the values or meaning that are generally associated to such a material component: the risks deriving from agribusiness, the industrialization and manipulation of food, and so on. Thus, if in the 1920–40s, when technological progress was emphasised, famous brands such as Barilla or Buitoni proudly advertised their "pastina glutinata" (gluten-enriched pasta), claiming that it helped healthy kids to grow, today gluten is rather condemned as the result of a manipulated and irresponsible science, which cannot but lead to disease and death (cf. Stano 2016).

In fact, this is part of a broader "praise for Nature" phenomenon, and the consequent condemnation of any "ultra-processing" of food (to recall the words of the Realfooding movement). Eating has been increasingly subjected to a logic of marked denial: to be considered healthy (that is, "good to think"), food must in a certain sense deny its own "culturalness", realizing a process that, drawing on Lotman's work, Cervelli (2012) has effectively described as an "incursion" of the extra-semiotic chaos of Nature (which is then positively connoted) into the semiotically ordered universe of Culture (dysphorised due to its presumptuous ambition to dominate Nature), as it is evident in the case of another crucial issue characterising contemporary foodscapes: the opposition between organic food, on one hand, and transgenic or genetically modified food, on the other hand.

11.4 Organic Versus GM Food: The Nature/Culture Myth

Although scientific consensus on the safety of GM foods has increased in recent years, most people still perceive them as a major threat for health. By contrast, the organic market is flourishing precisely for the opposite reason. While certainly linked to material factors, such trends also depend on specific mythologies—and, more precisely, on the above-mentioned diffused praise for "Nature" characterising contemporary foodspheres.

The World Health Organization defines GMOs as "organisms in which the genetic material (DNA) has been altered in a way that does not occur *naturally* by mating and/or *natural* recombination. The technology … allows selected individual genes to be transferred from one organism into another, also between nonrelated species" (WHO 2014, emphasis added).[7] Although largely shared and adopted by a number

paid to eating healthy food, Nicolosi criticises contemporary societies' obsession with food, as related not only to the fear of physical contamination, but also the fear of a loss of a symbolic-identity purity (Nicolosi 2006/2007: 49).

[7] Such a definition is echoed, with only slight variations, by other international institutions, such as the European Directive 2001/18/CE (https://eur-lex.europa.eu/legal-content/EN/TXT/?uri=CELEX%3A32001L0018), the American National Bioengineered Food Disclosure Standard (2018, https://www.federalregister.gov/documents/2018/12/21/2018-27283/national-bioengineered-food-disclosure-standard), the NON-GMO Project (https://www.nongmoproject.org), etc.

of institutional entities, such a description is problematic: not only does it refer to a classification system based on the techniques of production of the considered organisms rather than on their final characteristics, but it relies on manifestly arbitrary and not at all clear criteria for defining the "naturalness" of such techniques. According to the European legislation, for instance, some practices, such as the recombination of nucleic acids, cell fusion, or the injection of external heritable material into one organism, are considered forms of genetic modification, while others, such as the so-called mutagenesis or some methods of cell fusion not involving nucleic acids recombination, are not (cf. Bressanini and Mautino 2015; Stano 2021). This creates a series of borderline cases, as well as huge differences in legislation, with the US law HR 933 (2013) preventing federal judges from introducing any ban on the sale of GMOs (not without opposition and bottom-up initiatives increasingly manifesting discontent towards such a regulation[8]) versus Europe adopting a "precautionary principle" that bans any recognised genetic modification of food.

The same aura of indeterminateness surrounds organic food, whose definition also calls on nature: "as the chemical paradigm is about controlling *nature*, the organic paradigm is about respecting *nature*" (Rodale 2010, *Organic Manifesto*). However, what such a "nature" is is not always clear: sometimes defined as an "ecological balance" (SARE and the EU Commission), sometimes rather described in differential terms as "anything that is non-synthetic" (FAO and WHO), the term returns to emphasising the arbitrariness of the definition it is part of, pointing out the same issues characterising GM foods and their management.

From a semiotic point of view, it is particularly interesting to focus on the collective representations of the Nature/Culture mythology underlining present-day narratives concerning GM and organic food. First of all, it is essential to note the differential and inchoative characterisation of Nature suggested by such discourses: as shown in detail in Stano (2021), Nature is how things are "originally", with such an origin being either a "here" crystallised in an idyllic past to be preserved or rather a still uncontaminated "elsewhere" to be safeguarded. In any case, it consists in denying Culture, which is generally associated with a negative axiologisation. On the contrary, most representations of GM products are characterised as *terminative*, connoted as the end of a transformation process and even "subversion" of the natural order. This makes the mechanisms of manipulation of food, which are concealed as much as possible or at least disguised under the aesthetics of a distant tradition discursively associated with organic products, become predominant, finding expression through figurative elements such as the human hands (carefully protected by latex gloves as in any hazardous procedure) and injections, which are in fact recurrent.

Oranges-kiwis, apples-watermelons, banana peels revealing salamis inside them and other visual paradoxes, then, evidently mark a stigmatisation of such a process: if the lack of processes of alteration in organic products suggests a veridictory intent (that is, making such products *appear as they are*, cf. Greimas 1966), transgenic foods rather oscillate between secret (*they do not seem what they are*) and lie (*they are not*

[8] Especially as a result of food activism campaigns via social media, as highlighted by Adamoli (2012).

what they seem). Drawing on the distinction introduced by Floch (1990) between the representational and the constructive function of language, it is in fact possible to associate the organic market with a *referential* strategy based on a disengaged and informative style, often including quantitative data, graphs and nutritional tables to suggest that consumers are told the "truth" about products. Communication of GM foods, by contrast, seems to move between an *oblique* strategy based on paradoxes, hyperboles and forms of irony aimed at highlighting the dangers related to their consumption, and a *mythical* strategy placing food into new narratives and investing them with symbolic meanings. The recurrent figure of "Frankenfoods" is emblematic in this sense: pieces of fruit sutured together evidently evoke Mary Shelley's famous novel, thus recalling the dysphoric myth of a science that dares to go beyond the limits of humanity and is therefore destined to succumb to its own creations (cf. Ortoleva 2019). Such an idea is in fact very common in the debate on genetically modified foods. In 1998, for instance, Prince Charles reproached food biotechnologists for taking humankind into realms that belong exclusively to God, warning his readers about the "the long-term consequences for human health and the wider environment of releasing plants bred in this way" (Prince of Wales 1998). Similarly, in 2010 Pope John Paul claimed that using genetically modified organisms to increase production was "contrary to God's will", also advising farmers that when they "forget this basic principle and become tyrants of the earth rather than its custodians … sooner or later the earth rebels" (in Lyman 2000).

11.5 Conclusion

The brief overview examined here, while certainly capable of further development, is nonetheless significant, showing that, despite the peculiarities of each case, even in the field of nutrition which is generally thought to be regulated by "objective" and "measurable" material factors, materiality is always manifest in discourses. As such, it is given specific meanings and values, which inevitably affect the way we perceive it. More specifically, the nutritional dimension, which in itself is certainly regulated by materiality, becomes in such discourses a real system of signification of food—and, more specifically, a "system of classification" of edible and inedible substances, according to the crucial yet often unperceived processes described by scholars such as Douglas (1972) and Lévi-Strauss (see in particular 1962,1964).

This leads us to a final and crucial consideration, which points out a sort of "reversal" of course in food cosmologies—in Laudan's terms—, or mythologies—as we would say to emphasise Barthes' description of mythical language as a meta-language or operation of re-semantisation. From a system in which cooking was conceived, euphorically, as a bridge between (the disorder of) Nature and (the order of) Culture, according to the famous definition introduced by Lévi-Strauss (1965, 1966), modernity has marked first a shift to fermentation, that is, to a more natural transformation of food, and finally the passage to a real denial of Culture—a Culture that is no longer conceived as an Order, as it was in the systems studied by Douglas or Lévi-Strauss, but rather as an *insub-ordination* (to the laws of Nature); no longer as a form of science (in the Latin acceptation of *scientia*, that is to say, knowledge), but rather as a form of *incoscientia*, namely the dystopic realisation of a too pretentious humanity that does not recognise its limits. Only the future will show us where such myths will lead—provided that we succeed in the compelling and also necessary attempt, as this chapter has demonstrated, to deal with nutrition from a broader perspective, able to grasp the processes of meaning-making to which it is inevitably linked. It is in this sense—we might argue, paraphrasing a well-known quote by Lévi-Strauss (1965)—that we can hope to discover how nutrition is a (meta-)language in which societies unconsciously translate their structure, unless, equally unconsciously, they agree to reveal their contradictions.

References

Adamoli, Ginevra C.E.. 2012. *Social media and social movements: A critical analysis of audience's use of Facebook to advocate food activism offline (Doctoral dissertation).* Florida: College of Communication and Information, Florida State University.

Barthes, Roland. 1957. *Mythologies.* Paris: Editions de Seuil.

Barthes, Roland. 1961. Pour une psychosociologie de l'alimentation contemporaine. *Annales ESC* XVI(5): 977–986. English edition Barthes, Roland. 1997. Toward a Psychosociology of Contemporary Food Consumption. In *Food and Culture: A Reader*, ed. Carole Counihan and Penny Van Esterik, 20–27. New York and London: Routledge.

Bratman, Steven. 1997. The health food eating disorder. *Yoga Journal (September-October)*, 42–50.

Bratman, Steven, and David Knight. 2001. *Health food junkies. Orthorexia nervosa: Overcoming the obsession with healthful eating.* New York: Broadway.

Bressanini, Dario, and Beatrice Mautino. 2015. *Contro natura: Dagli OGM al "bio", falsi allarmi e verità nascoste del cibo che portiamo in tavola.* Milan: Rizzoli.

Cervelli, Pierluigi. 2012. Tecniche della natura umana. Biopolitica e semiotica del potere. In *Semiotica della natura (natura della semiotica)*, ed. Gianfranco Marrone, 103–117. Mimesis: Milan-Udine.

Douglas, Mary. 1966. *Purity and Danger. An Analysis of Concepts of Pollution and Taboo.* London: Routledge and Kegan Paul.

Douglas, Mary. 1972. Deciphering a meal. *Daedalus* 101(1): 61–81. Reprinted in Douglas, Mary. 1975. *Implicit Meanings*, 249–275. London: Routledge.

Elli, Luca, Federica Branchi, Carolina Tomba, Danilo Villalta, Lorenzo Norsa, Francesca Ferretti, Leda Roncoroni, and Maria Teresa Bardella. 2015. Diagnosis of gluten related disorders: Celiac disease, wheat allergy and non-celiac gluten sensitivity. *World Journal of Gastroenterology* 21 (23): 7110–7119.

Fasano, Alessio, Anna Sapone, Victor Zevallos, and Detleff Schuppan. 2015. Nonceliac gluten sensitivity. *Gastroenterology* 148 (6): 1195–1204.

Floch, Jean-Marie. 1990. *Sémiotique, marketing et communication.* Paris: PUF.

Greimas, Algirdas J. 1966. *Sémantique structurale.* Paris: Larousse. English edition Greimas, Algirdas J. 1983. *Structural Semantics: An Attempt at a Method* (trans: McDowell Daniele, Ronald Schleifer, and Alan Velie). Lincoln and London: University of Nebraska Press.

Greimas, Algirdas J., and Joseph Courtés. 1979. *Sémiotique: Dictionnaire raisonné de la théorie du langage.* Paris: Hachette.

Keys, Ancel. 1980. *Seven Countries. A multivariate analysis of death and coronary heart disease.* Cambridge and London: Harvard University Press.

Keys, Ancel, and Margaret Keys. 1975. *Eat Well and Stay Well, the Mediterranean Way*. Garden City: Doubleday.
Laudan, Rachel. 2004 [2000]. Birth of the Modern Diet. *Scientific American*, 283(2): 62–67.
Lévi-Strauss, Claude. 1962. *Le totémisme aujourd'hui*. Paris: PUF. English edition Lévi-Strauss, Claude. 1963. *Totemism* (trans: Rodney Needham). Boston: Beacon press.
Lévi-Strauss, Claude. 1964. *Mythologiques I. Le cru et le cuit*. Paris: Plon. English edition Lévi-Strauss, Claude. 1969. *The Raw and the Cooked* (trans: Weightman, John, and Doreen Weightman). Chicago: The University of Chicago Press.
Lévi-Strauss, Claude. 1965. Le triangle culinaire. *L'Arc* 26: 19–29. Reprinted in *Le Nouvel Observateur Hors-Série* (2009): 14–17. http://palimpsestes.fr/textes_philo/levi_strauss/triangle_culinaire.pdf. Accessed 30 April 2020. English edition Lévi-Strauss, Claude. 1997. The Culinary Triangle. In *Food and Culture: A Reader*, eds. Carole Counihan and Penny Van Esterik, 28–35. New York and London: Routledge.
Lévi-Strauss, Claude. 1966. *Mythologiques II. Du miel aux cendres*. Paris: Plon. English edition Lévi-Strauss, Claude. 1973. *From Honey to Ashes* (Translated by Weightman, John, and Doreen Weightman). New York: Harper and Row.
Ludvigsson, Jonas F., Daniel A. Leffler, Julio C. Bai, Federico Biagi, Alessio Fasano, Peter H.R.. Green, Marios Hadjivassiliou, Katri Kaukinen, Ciaran P. Kelly, Jonathan N. Leonard, Knut Erik Aslaksen. Lundin, Joseph A. Murray, David S. Sanders, Marjorie M. Walker, Fabiana Zingone, and Carolina Ciacci. 2013. The Oslo definitions for coeliac disease and related terms. *Gut* 62 (1): 43–52.
Lyman, Eric. 2000. Pope expresses opposition to GMOs: Cites need for 'the respect of nature'. *Bureau of National Affairs*, 221. http://online.sfsu.edu/~rone/GEessays/PopeGMO.htm. Accessed 14 April 2020.
Marrone, Gianfranco. 2014. *Gastromania*. Milan: Bompiani.
Marrone, Gianfranco. 2016. *Semiotica del gusto. Linguaggi della cucina, del cibo, della tavola.* Milan-Udine: Mimesis.
Marrone, Gianfranco. 2019. *Dopo la cena, allo stesso modo. Dieci anni di immaginario gastronomico*. Palermo: Torri del Vento.
Nestle, Marion. 2013 [2002]. *Food Politics: How the Food Industry Influences Nutrition, and Health*. Berkeley: University of California Press.
Nicolosi, Guido. 2006/2007. Biotechnologies, alimentary fears and the orthorexic society. *Tailoring Biotechnologies* 2(3): 37–56.
Nicolosi, Guido. 2007. *Lost food. Comunicazione e cibo nella società ortoressica*. Catania: Ed.it.
Niola, Marino. 2015. *Homo dieteticus. Viaggio nelle tribù alimentari*. Bologna: Il Mulino.
Ortoleva, Peppino. 2019. *Miti a bassa intensità. Racconti, media, vita quotidiana*. Turin: Einaudi.
Prince of Wales. 1998. Seeds of disaster. *Daily Telegraph* (London), June 8. Reprinted in *Ecologist*, 28(5): 252–253.
Schuppan, Detlef, Geethanjali Pickert, Muhammad Ashfaq-Khan, and VictorZevallos. 2015. Non-celiac wheat sensitivity: Differential diagnosis, triggers and implications. *Best Practice & Research Clinical Gastroenterology* 29(3): 469–476.
Scrinis, Gyorgy. 2008. On the ideology of nutritionism. *Gastronomica* 8 (1): 39–48.
Scrinis, Gyorgy. 2013. *Nutritionism: The Science and Politics of Dietary Advice*. New York: Columbia University Press.
Stano, Simona. 2016. Tell me what you do not eat, and i shall tell you what you are. Food, health, and conspiracy theories. *Lexia* 23–24: 327–343.
Stano, Simona. 2018. Glocal food and transnational identities: The case of the Mediterranean diet. In *Cross-Inter-Multi-Trans – Proceedings of the 13th World Congress of the International Association for Semiotic Studies (IASS/AIS)*, ed. Audra Daubariene, Simona Stano, and Ulrika Varankaite, 450–459. Kaunas: IASS Publications and International Semiotics Institute.
Stano, Simona. 2021. Food, Health and the Body: A Biosemiotic Approach to Contemporary Eating Habits. In *Food and Medicine: A Biosemiotic Perspective*, ed. Yogi H. Hendlin and Jonathan Hope, 43-60. Cham: Springer.

UNESCO. 2012. Mediterranean diet. http://www.unesco.org/culture/ich/en/RL/00884. Accessed 14 April 2020.

UNESCO. 2013. Mediterranean diet. http://www.unesco.org/culture/ich/RL/00884. Accessed 14 April 2020.

Volta, Umberto, Giacomo Caio, Francesco Tovoli, and Roberto De Giorgio. 2013. Non–celiac gluten sensitivity: Questions still to be answered despite increasing awareness. *Cellular and Molecular Immunology* 10 (5): 383–392.

Vriezinga, Sabine L., Joaquim J. Schweizer, Frits Koning, and M. Luisa Mearin. 2015. Coeliac disease and gluten-related disorders in childhood. *Nature Reviews Gastroenterology & Hepatology* 12 (9): 527–536.

WHO – World Health Organization. 2014. 3. Is the safety of GM foods assessed differently from conventional foods? Frequently asked questions on genetically modified foods. http://www.who.int/foodsafety/areas_work/food-technology/faq-genetically-modified-food/en/. Accessed 12 April 2020.

Simona Stano is Tenure-Track Assistant Professor at the University of Turin (UNITO, Italy) and Visiting Researcher at New York University (NYU, US). She has been awarded a Marie Curie Global Fellowship for a research project (COMFECTION, 2019–2021) on the semiotic analysis of food communication. She also worked as Senior Researcher at the International Semiotics Institute (2015–2018) and as Visiting Researcher at the University of Toronto (2013), the University of Barcelona (2015–2016) and Observatorio de la Alimentación (2015–2016).

She deals mainly with semiotics of culturefood semiotics, body semiotics, and communication studies, and has published several papers, edited volumes (including special issues of top semiotic journals such as *Semiotica* and *Lexia*), and monographs (*I sensi del cibo*, 2018; *Eating the Other. Translations of the Culinary Code,* 2015) on these topics. Since 2020, she is the vice-Director of the Centre for Interdisciplinary Research on Communication (CIRCe) of the University of Turin.

Chapter 12
Laughing Alone with Salad: Nutrition-Based Inequity in Women's Diet and Wellness Media

Emily J. H. Contois

Abstract Nutritionism creates hegemonic norms for eating and defining subjectivity, but this chapter demonstrates the specifically gendered effects of this ideology. Examining how diet and wellness media communicate stereotypical ideas about gender through salad, the chapter analyses a variety of source material, including the photo essay "Women Laughing Alone with Salad", the *Men's Health* and *Women's Health* websites, Instagram images, and numerous food and "wellness" trends. Such sources create nutritionism's subjects in ways closely aligned with existing dynamics of patriarchy, whiteness, affluence, and bodily ability and size. This analysis demonstrates how the refrain "salad is feminine" endorses and reinforces notions of identity, power, and inequity through specifically nutricentric means that permeate women's diet and wellness media.

Keywords Gender · Salad · Media · Nutritionism · Dieting

12.1 Introduction

Salad is feminine. Steak is masculine.

This is perhaps the most repeated adage expressing a gender binary through food. Despite its ubiquity, this dichotomy is neither essentially true nor natural. Rather, it is a food gender relationship constructed within and through culture, which historian Freedman (2019) argues emerged in American dietary advice, advertising, and magazines in the late-nineteenth century. Even so, what is supposedly feminine about salad? In her foundational work of feminist eco-criticism, *The Sexual PoliticsofMeat* (2015 [1990]), Carol J. Adams asserts that while meat connotes men, masculinity, and a powerful essence, vegetables have come to symbolize women, femininity, and passivity (14–15). Vegetables, such as those in salad, are not just passively feminine, but derisively so. Drawing from Hegel, Adams writes, "Women may eat plants, since each is placid, but active men need animal meat" (15–16).

E. J. H. Contois (✉)
Department of Media Studies, The University of Tulsa, Tulsa, OK, USA
e-mail: emily-contois@utulsa.edu

S. Stano and A. Bentley (eds.), *Food for Thought*, Numanities - Arts and Humanities in Progress 19, https://doi.org/10.1007/978-3-030-81115-0_12

One can also excavate salad's supposed and subjugated femininity through the ways that cultural meaning is embedded in nutrition education and research. Gyorgy Scrinis (2013) coined the concept of "nutritionism" to describe how nutrition science routinely focuses narrowly on nutrient components rather than on foods or foodways. Such studies also interpret these components' effects on bodily health in a reductive fashion. This perspective influences not only nutrition science research but also dietary policy and food marketing. In the case of salad, nutrition research, dietary advice, and food media alike promote leafy greens topped with colourful vegetables as a "healthy" meal option. Such sources focus on salad's nutritional components (such as fibre, antioxidants, vitamins, and low-calorie count), and these components' health benefits, such as satiety, weight loss or maintenance, digestive and bowel functionality, and cancer prevention.

Furthermore, Scrinis asserts that nutritionism constitutes an ideology that produces its own subjects: "nutricentric persons" (2013: 13). Nutritionism exerts outsized influence on how some eaters experience food and their bodies. For them, nutritional components and their associated health benefits guide all food decisions. These aspects significantly decentre the role of other aspects of food and eating, such as taste, pleasure, tradition, or seasonality, to name a few. Indeed, scholars and practitioners alike have made the call to "do nutrition differently" (Hayes-Conroy and Hayes-Conroy 2013). In her analysis of USDA federal food guides, Jessica Mudry (2010) similarly tracks how the language of quantification—one written in calories, grams of fat, and milligrams of sodium—rewrote many American eaters' relationship to food. For some, this has had deleterious consequences. She writes:

> Policy that employs this kind of language in attempts to rationalize the sensibilities of the human body is doomed to fail until it can supplement a discourse of quantification with alternative languages that speak to the subject of the body not the object of the body. (Ibid.: 3)

This statement takes on an even more urgent meaning when one considers not just universalised human bodies, but bodies differentiated by gender, race, social class, or citizenship status and related systems of domination and oppression. Women's bodies, in particular, have been misunderstood as irrational and women themselves as objects.

Indeed, when psychology scholar Paul Rozin (1999) famously wrote, "Food is fundamental, fun, frightening, and far-reaching" (9), he briefly explored how food's frightening aspects affect women disproportionately. Feminist philosopher Susan Bordo further analysed women's food fears in her foundational work, *Unbearable Weight* (2004 [1993]):

> Emotional heights, intensity, love, and thrills, it is women who habitually seek such experiences from food and who are most likely to be overwhelmed by their relationship to food, to find it dangerous and frightening (especially rich, fattening, soothing food like ice cream). (108)

In such ways, food's various capacities have different consequences for women. While sometimes a good deal less than fun, salads (especially of the type I analyse in this chapter) emerge as one of few foods deemed "safe", in large part on nutricentric terms based on aspirational nutritional components and the promise of low-calorie eating.

Existing dynamics of power and inequity further exacerbate the production of nutricentric persons. All nutricentric subjects are not similarly influenced (or oppressed) by nutritionism. This is especially the case as food media and popular culture spread nutritionism's views far and wide. This chapter explores how diet and wellness media communicate stereotypical ideas about gender through salad. These sources create nutritionism's subjects in ways closely aligned with existing dynamics of patriarchy, whiteness, affluence, and bodily ability and size. In this way, I demonstrate how the refrain "salad is feminine" endorses and reinforces notions of identity and systems of power and inequity through specifically nutricentric means.

12.2 Women Laughing Alone with Salad

Images that link salads and femininity abound across media forms, but in 2011, *The Hairpin* founder, Edith Zimmerman, published the digital photo essay "Women Laughing Alone with Salad".[1] Published in January of that year, Zimmerman's ironic piece coincided with the moment when the weight loss industry's "new year, new you" messaging was at its seasonal height. In the essay's images, women—mostly young, white, and thin—sit or stand alone with their salad. They laugh with their mouths open wide to show their perfectly straight white teeth and to express their exaggerated joy from consuming salad. I have taught this essay in a number of college classrooms and presented it to various groups outside of academia as well. It always draws laughter, but from women in the audience, it also draws a different note of recognition and frustration. The photo essay hit a nerve, as it inspired memes and even a well-received theatrical play (Know Your Meme 2016; Gray 2016).

Then as now, the essay communicated a number of gendered critiques. These images of gloriously happy, salad-eating women revealed the state of stock photography, which often depicts women in limited and stereotypical ways (Frosh 2001; McCarthy 2015). Such images of women and salad frequently accompany health promotion literature and online "healthy eating" content. In other fields, such as business, stock photography long offered only shots of secretaries or women dressed in suits juggling babies (Cunningham 2014). Getty Images launched the Lean In Collection of images in 2014 to address this problem and to provide "more empowering" images of women, particularly in leadership roles. But the "Women Laughing Alone with Salad" images demonstrate how representations of food and eating index gender in more complex ways than simply stereotypical images of secretaries in cat eye glasses.

First, these stereotypes are not just visual. The essay title offers the only text of the essay, and its phrasing is worthy of consideration. It follows the naming convention of

---

[1] *The Hairpin* was a women's general interest website. It ceased publication at the end of January 2018.

artworks, particularly still life pieces. Consider Pieter Claesz's "Vanitas with Violin and Glass Ball" (1628) or Francisco de Zurbaran's "Still Life with Lemons, Oranges and a Rose" (1633). In Zimmerman's case, women are the primary object of the still life. This naming convention grants them action but only through forced laugher. These women are alone except for salad, the object which they are "with".

In this way, Zimmerman textually and visually critiqued the objectification of women. Indeed, the women depicted in these images are objects rather than subjects. A few women look directly into the camera, but many, if not most, lack any direct gaze. Instead, these women lift their chins skyward as they laugh, hard, at the joy of eating salad. Others look downward through their eyelashes, bashfully and thoughtfully, at their vegetables.

In a few such images, salad plates sit on tables as if for a meal. Far more often, a woman holds the salad dish or bowl mid-air, cupped in her hand. As petite as these women's bodies, these salad-holding vessels occupy a different space with less permanence, less authority, less realness than a hefty, hot plate loaded with steak and potatoes hunkered down on a table. These images often depict forks pierced with colourful bites, but hardly ever show any actual eating. Forks serve as functional props for these staged performances of happy, healthy consumption and of white straight femininity, both of which these images present as aspirational, ideal types. These forks shorten the distance between salads and mouths, food and sustenance, but never fulfil their purpose. In such ways, these images reinforce women's objectification and limit their subjectivity through a number of visual and textual means, all linked to a supposedly feminine food: salad.

12.3 Sad Salads: Mixing Negative Calorie Lore with Lack and Normative Discontent

These images also depict salads in specific and constrained ways. They are what one of my students once called "sad salads". Not particularly verdant or inspiring, these salads are combinations of mostly lettuce and water-filled vegetables, such as cucumbers, tomatoes, and colourful pepper slices. These are salads without tasty and balanced accompaniment, such as cheese, nuts, or creamy dressings, although one can sometimes glimpse a chickpea or two.

These salads endorse the wholly nutricentric diet message of not just low-calorie eating, but the lore of "negative calories"—the idea that some high-fibre and high-water foods supposedly "burn" enough calories through chewing, digestion, metabolism, and excretion to render a caloric deficit and weight loss (Buddemeyer et al. 2019). Sad salad photos often depict hallmark negative-calorie foods: lettuce, celery, cucumber, and broccoli. These images also bespeak an obsession with the calorie itself and its tight control, which has historical, political, and economic consequences (Cullather 2007; Nestle and Nesheim 2012). In the United States, much public health anti-obesity messaging endorses the notion of energy balance, described in terms of "calories in" and "calories out". The advice for maintaining weight is to burn in a day the same number of calories that one eats, or to take in fewer calories than one burns in order to lose weight. Negative calorie lore takes such recommendations to an illusory next step, seeking to find in food itself—especially foods like salad—a source of desired energy imbalance.

Despite being repeatedly debunked (O'Connor 2006; Wilson 2019; Buddemeyer et al. 2019), negative calorie lore still infiltrates food media. For example, the article "Negative Calorie Foods: You Can Eat These 11 Foods & Not Gain Weight" was updated in 2018, but still endorses the negative-calorie approach. Diet books also promote this unfounded weight loss advice, such as *The Negative Calorie Diet* (DiSpirito 2015). On Instagram in spring 2020, the hashtag #negativecalories had drawn approximately 2,800 posts. Its unpopularity is notable, especially when compared to somewhat similar hashtags like #cleaneating (46.3 million posts), #detox (17 million posts), or #whole30 (4.3 million posts). A handful of accounts seem to contribute the majority of posts on negative-calorie eating. @crazyketoguy, who uses his website and Instagram account to track his weight loss, affixes the hashtag to all of his posts, which depict meals, workouts, and snapshots of glucose monitors. @fitTox, described as "India's first Ayurvedic Detox Teas & Coffee", posted an image of whimsical vegetable figures, writing in the caption, "Eating these vegetables, you burn more calories than what they add to your body". The Italian health and wellness account @bene_mangia_sano, boasts more than 54,000 followers and frequently posts about "dieta a calorie negative". Three of the posts in the #negativecalories top nine in April 2020 were images of colourful fruits and vegetables from this account.

Across food media and cultures, negative calorie lore has drawn peripheral but sustained attention for its falsely promised weight loss effects. Taking gender into account, sad salads and negative calorie lore are further linked by the expectation for female lack. Social conventions require women to pursue thinness, control their appetites, take up less space, and embody a state of lack (Bordo 2004 [1993]; Hartshorne 2014). Related to Adams' claims earlier regarding femininity, plants, and passivity, Irigaray (1985) somewhat similarly theorised women's sexuality, sexual pleasure, and agency, as she built from Freud's theory of penis envy. When "conceptualized on the basis of masculine parameters", she writes, women's sexuality "has nothing to say", as "her lot is that of lack", given the supposed femininity and passivity of the vagina, compared to "the noble phallic organ" (Ibid.: 23).

Social conventions seek to constrain women's appetites for both sex and food, demanding chaste morality in and through "good" eating and foods, like sad salads. As a result, social expectations for female thinness cause unending feelings for women that one's body is not good enough. Psychologists Judith Rodin, Lisa Silberstein, and Ruth Striegel-Moore (1984) coined the term "normative discontent" to describe the widespread dissatisfaction that most women feel about their bodies.

"Women Laughing Alone with Salad" joins the forces of negative calorie lore and normative discontent, fusing them together into a nonsensical *embrace* of lack. These women do not just consume salad, they do so joyfully, in a state of nearly orgasmic *un*consumption. They are *fulfilled* by foods that provide relatively little sustenance—or in negative-calorie lore, foods that offer an elusive and desirable negativity. These images depict the most complete indoctrination of normative discontent possible, as the pursuit of lack is not just normal, but expanded into a state of exaggerated joy and pleasure, expressed through open-mouthed laughter.

These images also bring to life the complicated dynamics of health that Charlotte Biltekoff analysed in *Eating Right in America* (2013). Eaters are not just encouraged to follow dietary advice, she argues, but to embody it. She writes that the good food movement advised eaters to not just change their behaviours and eat foods like kale, but to become the sort of person who *likes* kale, the sort of person who *desires* kale (Ibid.: 89). Even among the generally lauded leafy greens of salads, kale rose to nutritional prominence in the 2010s for its significant vitamin and mineral content. Nutrition reporting online hails kale for its significant amounts of vitamin C, A, and K, calcium, and omega3 fatty acid (Carruthers 2018; Glassman no date). Given the cultural pressures of the thin ideal, nutritional admonishment carries an even heavier weight for female eaters.

Demonstrating further gendered meanings, the imperative to desire kale collides with commodity feminism (Gill 2008; Banet-Weiser 2018) in fascinating and problematic ways. Kale transforms into a fashion statement when emblazoned onto organic cotton t-shirts and coffee mugs. A certain type of "enlightened" consumer embodies health as she desires and eats kale, bringing it *into* her body. She also purchases, wears, and displays representations of kale *on* her body. Despite the popular proliferation of kale, the initial attractiveness of this leafy green remains rooted in nutrition and on nutricentric terms, so long endorsed by sad salads.

12.4 From Sad Salads to Power Bowls: Gender in the Shift from Diet to Wellness

Kale or otherwise, stock images certainly exist of men laughing while eating salads, but there are fewer such images than those featuring women. Furthermore, these images of men do not co-exist within the same volume of "sad salad" nutricentric advice as that aimed at women. Salads intended for men and male consumption are not sad salads nor do they represent quests for negative-calorie eating. While gendered conventions mandate restrained appetites for women, salads for men build upon notions that embrace male appetites as voracious, hearty, and requiring satisfaction (Kiefer et al. 2005; Wardle et al. 2004).

Such conventions come to life when comparing, for example, the search results for "salad" on the *Men's Health* versus *Women's Health* websites, which I performed in April 2020. The search drew 50 percent more results for women, 124 versus 82 for men. Women's salad articles and recipes addressed topics such as tasty low-calorie dressings, low-calorie salads generally, and recipes to fight salad fatigue. Such salad articles even included "The Vegan Food List You Need to Take on Your

Next Grocery Trip" (Yu 2019), which promised "no more sad salad", acknowledging that *Women's Health* readers have likely eaten a fair number of them. In comparison, men's salad articles distanced themselves from calories and dieting altogether. Such titles included, "15 Hearty Keto Salads That Won't Make You Feel Like You're on a Diet" (Peters 2019) and "6 High-Protein Foods to Add to Your Salad" with the teaser copy, "Salads are boring. But they don't have to be" (Kadey 2018). These men's salad recipes also included the "Big Ass Breakfast Salad with Bacon Dressing" (Matthews 2018), which incorporated playfully profane masculine bravado with the most un-sad-salad ingredient of all: bacon.

As this very brief comparison shows, content addressing women's and men's salads both adopt messaging focused on nutrients, such as calories, protein, and vitamins, but in divergent ways. Such differences reveal how nutritionism variably creates subjects along the lines of gender. In this way, women's salads (and "diet" foods more broadly) have long emphasised what they lack (such as calories or fat) and emphasised a conflicted appetite that women restrain in the pursuit of a thin ideal. Men's salads promote what they contain (such as protein and various vitamins) and emphasise a voracious appetite that deserves satisfaction in the pursuit of a muscular ideal. I call this the concept of "zero"—that is, a masculine substitute for the feminised notion of "diet" (Contois 2020).

Such gendered salad dichotomies are admittedly, and thankfully, starting to break down. Even within the current websites for *Women's Health* and *Men's Health*, women's salads also emphasise power, protein, satiety, and tasty options—messaging previously reserved for mostly male audiences. Indeed, while protein messaging typically dominated men's health advice, in the 2010s it drew attention from every corner of the food industry, stretching to encapsulate even food for babies and pets (Molyneaux 2015; Contois 2019). This protein mania reconfigured salads in the food marketplace into a new menu offering: "power bowls". A number of fast casual restaurants with varying price points started offering power bowls in the mid-to-late 2010s, such as Sweetgreen, Chop-T, Starbucks, Panera Bread, Chipotle, and Panda Express, to name but a few. Blogs, Pinterest boards, and Instagram also serve up numerous recipes for power bowls following the formula of grains, greens, and protein. While maintaining the appearance of salads, power bowls resignified such items, but maintained a nutricentric focus.

Beyond cultural protein obsession, these power bowl developments also related to larger cultural shifts from overt dieting to "wellness". In the later 2010s, efforts to lose (or otherwise manage) weight and to optimise health adopted a new language, which furthered past attempts to rebrand dieting as "a healthy lifestyle". Consumers lost interest in dieting and weight loss, seeking instead to be healthy and fit, clean and strong, empowered and feminist (Brodesser-Akner 2017). But female consumers still desired (or felt pressured to achieve) thinness. Wellness simply shrouded dieting with a shiny new veneer (Fabello 2019; Knoll 2019). It marked a change in vernacular more than a paradigm shift. While wellness messaging more rarely mentions calories directly (which are now viewed as vestiges of a passé diet mindset), it still endorses scientific reductionism and interpretation for food's components.

To return to Scrinis' (2013) work on nutritionism, he asserts that our food culture has been infiltrated by functional foods and perspectives. By this logic, many eaters, especially those subscribed to a wellness perspective, consume foods not to fulfil bodily needs, but to *optimise* health, to be better, to become more than. While wellness appears on the surface to have scraped away the pursuit of unreasonable and out-of-reach ideals for women, such as thinness, it continues to perpetuate the pursuit of optimisation, whether a maximised microbiome, an amplified immune system, or perfectly glowing skin.

Rife with such contradictions, the wellness shift rewrote the cultural meaning of salads, but in highly ambivalent ways. Within the purview of power, wellness, and optimisation, salads are no longer sad. They are no longer stripped down to their minimum caloric possibility. Salads for both women and men now adopt the "zero" perspective, focusing on salads that are bursting full of "superfood" additions, such as quinoa or other so-called "ancient grains", along with kale, chia seeds, and fermented toppings (Loyer and Knight 2018; Sikka 2019). While no longer emphasising lack, these salads still maintain a narrow and reductive focus that lauds salad components less for their taste, pleasure, or beauty, and more for their nutritional and health benefits.

Beyond their focus on wellness and strength, today's power bowl salads also seek to communicate "realness". Diet food (which is for women by default) is rarely real food (Contois 2017). The sad salads Zimmerman captured in her photo essay largely echoed the feminised diet food landscape of the 1990s and early 2000s. Grocery aisles offered a plethora of decidedly diet, very sweet, and overtly fake food options that first sprouted in the 1960s (de la Peña (now Thomas) 2010). Take for example Yoplait light yogurt, sugar-free Jell-O, SnackWell's cookies, or Special K cereals—all artificially sweetened foods meant to substitute for the real thing with as few calories as possible.

Within such a context, salad as a diet food was somewhat exceptional for its "real" ingredients. The "fakeness" usually congregated in the low- or zero-calorie dressings. Such an approach to eating bears both similarities and differences with today's "whole food" and protein-focused approach. The wellness shift remains constrained, however. For example, zucchini noodles or cauliflower pizza may be made with "real" ingredients, but they similarly substitute for their carbohydrate-filled originals. Wellness marches forward as dieting by another name, still promoting highly monitored consumption, including in power bowls.

Salad media, such as that on the *Women's Health* website, place new and welcome emphasis on taste, fulfilment, and strength. But even so, these salads remain wrapped up in the pursuit of a thin, fit ideal that has yet to be fully extricated from wellness. The emphasis upon power and protein for women brings about a degree of empowerment and gender equity in food media and marketing. At the same time, this obsession with protein brings compromises for our global food system, marked by insecurity and inequity. The availability of high protein salads for women in the global north still maintains global inequalities of access and sustainability. In such ways, the nutricentric meaning-making of salads and power bowls continues to propagate imbalances of power.

12.5 Conclusion: Why "Salad is Feminine" Matters

This chapter explored the role of identity and power within the ideology of nutritionism, paying particular attention to the dynamics of gender. Nutritionism focuses primarily on food's nutritional components. As a result, it creates a host of nutricentric persons, but they do not all move through the world of food equally. A close examination of the notion that "salad is feminine" and the photo essay, "Women Laughing Alone with Salad", demonstrates how nutritionism functions within current dynamics of power, particularly when it comes to gender and the subjugation of femininity and women, as well as race, class, and body shape and size. When viewed as solely calories, fibre, and vitamins, salad (and the *meaning* of salad circulated in media and culture) serves to propagate hierarchical notions of restraint, control, discontent, and lack. Even when reconfigured into power bowls and situated within a context of anti-diet wellness, salads today continue to communicate ideas about gender that are directly linked to nutrition-oriented ways of thinking. Such developments reveal the ongoing gaps within a nutrition and media future that promotes the health of everyone—one in which we might all eat, and laugh, together.

References

Adams, Carol. 2015 [1990]. *The Sexual Politics of Meat: A Feminist-Vegetarian Critical Theory*, 25th Anniversary Edition. New York and London: Bloomsbury Academic.

Banet-Weiser, Sarah. 2018. *Empowered: Popular Feminism and Popular Misogyny*. Durham: Duke University Press.

Biltekoff, Charlotte. 2013. *Eating Right in America: The Cultural Politics of Food and Health*. Durham: Duke University Press.

Bordo, Susan. 2004 [1993]. *Unbearable Weight: Feminism, Western Culture, and the Body*, Tenth Anniversary Edition. Oakland: University of California Press.

Brodesser-Akner, Taffy. 2017. Losing it in the anti-dieting age. *New York Times*, August 2. https://www.nytimes.com/2017/08/02/magazine/weight-watchers-oprah-losing-it-in-the-anti-dieting-age.html. Accessed 28 April 2020.

Buddemeyer, Katherine M., Ashley E. Alexander, and Stephen M. Secor. 2019. Negative calorie foods: An empirical examination of what is fact or fiction. *bioRxiv*, March 24. https://doi.org/10.1101/586958.

Cullather, Nick. 2007. The foreign policy of the calorie. *The American Historical Review* 112 (2): 337–364.

Carruthers, Linda. 2018. The many types and health benefits of kale. *Mayo Clinic*, May 29. https://www.mayoclinichealthsystem.org/hometown-health/speaking-of-health/the-many-types-and-health-benefits-of-kale. Accessed 28 April 2020.

Contois, Emily. 2017. Real men & real food: The cultural politics of male weight loss. *Nursing Clio*, August 15. https://nursingclio.org/2017/08/15/real-men-real-food-the-cultural-politics-of-male-weight-loss/. Accessed 28 April 2020.

Contois, Emily. 2019. Protein in the macronutrient imaginary: The case of 'brogurt' marketing. *H-Nutrition*, September 10. https://networks.h-net.org/node/134048/discussions/4443764/protein-macronutrient-imaginary-case-%E2%80%9Cbrogurt%E2%80%9D. Accessed 28 April 2020.

Contois, Emily. 2020. *Diners, Dudes, and Diets: How Gender and Power Collide in Food Media and Culture*. Chapel Hill: University of North Carolina Press.

Cunningham, Lillian. 2014. Changing our image of women, one stock photo at a time. *Washington Post*, February 10. https://www.washingtonpost.com/news/on-leadership/wp/2014/02/10/changing-our-image-of-women-one-stock-photo-at-a-time/. Accessed 28 April 2020.
de la Peña (now Thomas), Carolyn. 2010. *Empty Pleasures: The Story of Artificial Sweeteners from Saccharin to Splenda*. Chapel Hill: The University of North Carolina Press.
DiSpirito, Rocco. 2015. *The Negative Calorie Diet*. New York: Harper Collins.
Fabello, Melissa. 2019. "Is your wellness practice just a diet in disguise? *Bon Appetit*, April 3. https://www.bonappetit.com/story/is-your-wellness-practice-just-a-diet-in-disguise. Accessed 28 April 2020.
Freedman, Paul. 2019. How steak became manly and salads became feminine. *The Conversation*, October 24. https://theconversation.com/how-steak-became-manly-and-salads-became-feminine-124147.
Frosh, Paul. 2001. Inside the image factory: Stock photography and cultural production. *Media, Culture & Society* 23 (5): 625–646.
Gill, Rosalind. 2008. Commodity feminism. *The International Encyclopedia of Communication*. https://doi.org/10.1002/9781405186407.wbiecc055.
Glassman, Keri. No date. Why is kale good for me? *WebMD*. https://www.webmd.com/food-recipes/features/why-is-kale-good-for-me#1. Accessed 28 April 2020.
Gray, Margaret. 2016. Critic's choice: 'Women laughing alone with salad': A playwright's raucous riff on sexism and body shame. *Los Angeles Times*, March 14. https://www.latimes.com/entertainment/arts/la-et-cm-women-laughing-alone-with-salad-review-20160314-story.html. Accessed 28 April 2020.
Hartshorne, Sarah. 2014. I was a woman laughing alone with salad, it's really not that funny. *The Guardian*, March 5. https://www.theguardian.com/women-in-leadership/2014/mar/05/woman-laughing-alone-with-salad. Accessed 28 April 2020.
Hayes-Conroy, Allison, and Jessica Hayes-Conroy, eds. 2013. *Doing Nutrition Differently: Critical Approaches to Diet and Dietary Intervention*. Abingdon and New York: Routledge.
Irigaray, Luce. 1985. *This Sex Which Is Not One*. Translated by Catherine Porter with Carolyn Burke. Ithaca: Cornell University Press.
Kadey, Matthew. 2018. 6 High-protein foods to add to your salad. *Men's Health*, October 2. https://www.menshealth.com/nutrition/a23572071/high-protein-salad/. Accessed 28 April 2020.
Kiefer, Ingrid, Theres Rathmanner, and Michael Kunze. 2005. Eating and dieting differences in men and women. *Journal of Men's Health and Gender* 2 (2): 194–201.
Knoll, Jessica. 2019. Smash the wellness industry. *New York Times*, June 8. https://www.nytimes.com/2019/06/08/opinion/sunday/women-dieting-wellness.html. Accessed 28 April 2020.
Know Your Meme. Added in 2016, updated in 2018. Women laughing alone with salad. https://knowyourmeme.com/memes/women-laughing-alone-with-salad. Accessed 28 April 2020.
Loyer, Jessica, and Christine Knight. 2018. Selling the 'Inca superfood': Nutritional primitivism in superfoods books and maca marketing. *Food, Culture & Society* 21 (4): 449–467.
Matthews, Melissa. 2018. Big ass breakfast salad with bacon dressing. *Men's Health*, December 11. https://www.menshealth.com/fitness/a25471581/big-ass-breakfast-salad-with-bacon-dressing/. Accessed 28 April 2020.
McCarthy, Justin. 2015. Women laughing alone with salad: The meme, the myth, the legend. *Medium*, August 5. https://medium.com/collective-rage-a-play-in-five-boops/women-laughing-alone-with-salad-the-meme-the-myth-the-legend-88b7179689ea. Accessed 28 April 2020.
Molyneaux, Maryellen. 2015. Consumer protein trends. *Natural Products Insider*, October 8. http://www.naturalproductsinsider.com/articles/2015/10/consumer-protein-trends.aspx. Accessed 28 April 2020.
Mudry, Jessica J. 2010. *Measured Meals: Nutrition in America*. Albany: State University of New York Press.
Nestle, Marion, and Malden Nesheim. 2012. *Why Calories Count: From Science to Politics*. Oakland: University of California Press.

O'Connor, Anahad. 2006. The claim: Some foods have negative calories. *New York Times*. July 25. https://www.nytimes.com/2006/07/25/health/25real.html. Accessed 28 April 2020.
Peters, Jolie. 2019. 15 hearty keto salads that won't make you feel like you're on a diet. *Men's Health*. September 8. https://www.menshealth.com/nutrition/g28956027/hearty-keto-friendly-salads/. Accessed 28 April 2020.
Rodin, Judith, Lisa Silberstein, and Ruth Striegel-Moore. 1984. Women and weight: A normative discontent. *Nebraska Symposium on Motivation* 32: 267–307.
Rozin, Paul. 1999. Food is fundamental, fun, frightening, and far-reaching. *Social Research* 66 (1): 9–30.
Scrinis, Gyorgy. 2013. *Nutritionism: The Science and Politics of Dietary Advice*. Columbia University Press.
Sengupta, Sushmita. 2018. Negative calorie foods: You can eat these 11 foods & not gain weight. *NDTV Food*, updated August 22. https://food.ndtv.com/food-drinks/11-foods-that-burn-more-calories-than-they-contain-1679965. Accessed 28 April 2020.
Sikka, Tina. 2019. The contradictions of a superfood consumerism in a postfeminist, neoliberal world. *Food, Culture & Society* 22 (3): 354–375.
Wardle, Jane, Anne M. Haase, Andrew Steptoe, Maream Nillapun, Kiriboon Jonwutiwes, and France Bellisle. 2004. Gender differences in food choice: The contribution of health beliefs and dieting. *Annals of Behavioral Medicine* 27 (2): 107–116.
Wilson, Christy. 2019. 'Negative-calorie foods' still count. *Eat Right: Academy of Nutrition and Dietetics*, July 22. https://www.eatright.org/health/weight-loss/fad-diets/negative-calorie-foods-still-count. Accessed 28 April 2020.
Yu, Christine. 2019. The vegan food list you need to take on your next grocery trip. *Women's Health*, July 11. https://www.womenshealthmag.com/food/a28322972/vegan-food-list/. Accessed 28 April 2020.
Zimmerman, Edith. 2011. Women laughing alone with salad. *The Hairpin*, January 3. https://www.thehairpin.com/2011/01/women-laughing-alone-with-salad/. Accessed 28 April 2020.

Emily J. H. Contois is Assistant Professor of Media Studies. She researches, teaches, and writes on food, identity, and health in U.S. popular culture and media. In addition to articles in *Gastronomica, Feminist Media Studies*, and *American Studies*, she is the author of *Diners, Dudes, and Diets: How Gender and Power Collide in Food Mediaand Culture* (UNC Press 2020) and co-editor of Food Instagram: Identity, Influence, and Negotiation (University of Illinois Press, 2022). She is the Book Reviews Editor for *Food Culture & Society* and serves on the boards of the Association for the Study of Food and Society, H-Nutrition (for the history of nutrition science), and the Bloomsbury Food Library. Dedicated to public scholarship, she writes for *Nursing Clio*, blogs at emilycontois.com, and is active on social media at @emilycontois.

Chapter 13
Virtue and Disease: Narrative Accounts of Orthorexia Nervosa

Lauren A. Wynne, Gareen Hamalian, and Neve Durrwachter

Abstract Orthorexia nervosa is most frequently described as a fixation with healthy eating. Although orthorexia nervosa (ON) has yet to be added to the Diagnostic and Statistical Manual of Mental Disorders (DSM) and much debate persists as to its prevalence and criteria for its diagnosis, many mental health clinicians have observed the growing prevalence of the condition. However, the contemporary imperative in the United States to "care for" one's body through food, which includes the employment of nutritional knowledge and often the identification of "good" and "bad" foods, the diagnostic criteria for orthorexia reflect what are, to some degree, normalised, even valorised, approaches to food in American society. This chapter explores contestation over the production of orthorexia as a disease within medical and lay discourses. This essay also considers how self-identified recovering orthorexics make use of both the contested diagnosis criteria and the contested condition itself in narrating the trajectories of their experiences. Given that ON's existence is still debated and that solid empirical data on its prevalence is lacking, these authors play an important role in normalising the notion that healthful eating can become unhealthy.

Keywords Orthorexia nervosa · Eating disorders · Nutrition · Healthism · Medicalisation

---

L. A. Wynne (✉) · N. Durrwachter
Department of Anthropology and Sociology, Ursinus College, Collegeville, PA, USA
e-mail: lwynne@ursinus.edu

N. Durrwachter
e-mail: ndurrwa1@jh.edu

G. Hamalian
Department of Psychiatry, New York University, New York, NY, USA
e-mail: Gareen.Hamalian@nyulangone.org

S. Stano and A. Bentley (eds.), *Food for Thought*, Numanities - Arts and Humanities in Progress 19, https://doi.org/10.1007/978-3-030-81115-0_13

13.1 Introduction

"A disease disguised as a virtue", Bratman and Knight (2000) wrote of orthorexia nervosa in 2000. Orthorexia nervosa (ON) is most frequently described as a fixation with "healthy" eating; sufferers are thought to hold "magical beliefs" about the properties of certain foods and "intense phobia" towards other less-healthy foods (Lindeman et al. 2000; Koven and Abry 2015). Unlike other eating disorders, its ostensible motivation is not a particular bodily presentation or weight but rather healthfulness. Many clinicians note the growing prevalence of the condition, but there is little agreement even among medical providers on the best way to identify the disease. Furthermore, adherents of more restrictive diets, including some members of the medical community, resist what they perceive as the medicalisation of their food choices. However, growing attention to the condition on the part of self-identified recovering orthorexics has propelled the condition into popular discourse, even before it has been formalized within the Diagnostic and Statistical Manual of Mental Disorders (DSM). The contemporary proliferation of diet-related information, often pointed to as a causal factor in the prevalence of ON, has also become a force for the reification of the condition.

This chapter examines debates about ON, first and briefly, within the medical literature that is attempting to develop diagnostic criteria and better understand the condition, and, second, through analysis of responses to orthorexia outside of formal medical scholarship, through popular and social media. The juxtaposition of the contested medical literature with narrative accounts by individuals who identify themselves as recovering orthorexics reveals how ON is becoming a salient disease category without even yet being an entry in the DSM.

13.2 The Birth of a Disease

Physician and self-described recovering orthorexic Steven Bratman coined the name of the disease in a 1997 article in *Yoga Journal* (Bratman 1997). He later expanded this work in a book published in 2000, exploring both his own experience as an orthorexic and his experience diagnosing and treating the condition. It was not until 2004 that the condition was examined by a peer-reviewed journal, in this case in an article based on research conducted in Italy (Donini et al. 2004). By the end of that decade, the condition had attracted significant media attention. In the last five years, hundreds of media outlets have reported on the condition, and there has been a small but growing body of English-language scholarship on the disease. A recent Google Scholar search found over two thousand scholarly sources in over a dozen languages.

Much of the original research circulating on orthorexia featured case studies in which the authors described their attempts to identify and treat the pathologies presented in clinical settings. What differentiated these cases from those of individuals diagnosed with other eating disorders was the presence of "good health" as a motivating factor for dietary choices. Beyond that, however, the individual patients examined presented a range of symptoms, from the physiological—evidence of malnutrition—to the psychosocial—fractured relationships or financial challenges due to the amount of time or money devoted to food preparation and consumption. In their work on the condition, in order to reflect this range of symptomology, a trio of psychologists describe ON as potentially "a multidimensional construct" (Gleaves et al. 2013: 14). Similarly, while the scholarly literature now includes multiple means of assessing orthorexic symptomatology, there is little consensus on which assessments are best. These assessments range from Bratman's informal 10-item questionnaire to a more complex set of criteria published by Moroze et al. in (2015), Bratman and Knight (2000), Moroze et al. (2015). Bratman and Dunn (2016) later built on both of those to develop another set of criteria. A questionnaire developed a few years earlier, known as ORTO-15, included some of Bratman's original ten questions, plus some additional questions (Donini et al. 2005); this diagnostic tool has since been modified by other researchers and employed in multiple studies (e.g. Ramacciotti et al. 2011; Alvarenga et al. 2012; Varga et al. 2014). However the use of ORTO-15 has been critiqued in recent years, as some of those studies have reported rates of as high as 88.7% of subjects meeting the diagnostic criteria (Missbach et al. 2015; Dunn and Bratman 2016).[1]

At this time, however, orthorexia nervosa (ON) has yet to be added to DSM, in part because there is still not clear agreement as to the appropriate criteria for diagnosis or the particular ways in which the condition may overlap with other psychiatric disorders. Published by the American Psychiatric Association (APA), the DSM has served as a guide for diagnosing and coding psychiatric disorders since 1952 (Kawa and Giordano 2012). It is periodically revised through a process of research and decision-making on the part of APA-appointed committees of experts (APA 2011; Greenberg 2013). Now in its fifth iteration, the text relies on symptom-based criteria to standardize disease categories, which then receive codes, something insurance companies have come to rely on for determining reimbursement. Furthermore, a majority of committee members tasked with revising the two most recent DSM have reported ties to pharmaceutical companies (Cosgrove and Krimsky 2012). The addition of orthorexia nervosa to the next DSM would not then be a simple matter of recognizing and formalizing a disease, but rather might be better described as a politically and economically-laden process of disease creation, albeit one that may validate and propose to ease real forms of suffering.

The conceptual and practical power of the DSM lies in the contemporary extension of its applications outside of individual clinical spaces. While there seems to be a public misconception of the DSM as the "Bible" of psychiatry, those who work in the field recognize its limitations and are more apt to think of it as a field guide. For clinicians, the DSM offers a practical way of tying bundles of symptoms to possible illness trajectories and responses to treatments. In the preface of the DSM 5, the

[1] See Dunn and Bratman (2016) for an exhaustive review of the literature on ON. Håman et al. (2015) offer an excellent integrative literature review. We have opted to provide only a few key points to trace the emergence of and contestation about the condition.

authors present the book as "intended to serve as a practical, functional, and flexible guide for organizing information that can aid in the accurate diagnosis and treatment of mental disorders" (APA 2013: Preface). The authors are forthcoming about limitations: "Since a complete description of the underlying pathological processes is not possible for most mental disorders, it is important to emphasize that the current diagnostic criteria are the best available description of how mental disorders are expressed and can be recognized by trained clinicians" (Ibid.). Given the limited success in connecting mental health disorders to specific neurobiological markers, the DSM "gives clinicians a common language to describe and treat mental disorders" until the underlying biology of these disorders are better understood, if they ever are (Friedman 2013).

Importantly, ON is also not yet in the International Classification of Diseases (ICD), the WHO-published handbook and the DSM's main competitor in standardizing mental health disorders. But, in the American medical community, the DSM remains more influential in determining diagnosis and insurance reimbursement. For both manuals, the formation of an Orthorexia Nervosa Task Force in 2016, comprised almost entirely of researchers from the United States and Europe, suggests that ON is a strong contender for inclusion in one or both, either as a stand-alone category or as a variant of an already-recognized disorder (Cena et al. 2018).

13.3 ON and Contemporary Eating

In her philosophical analysis of ON, van Dyke (2018) reminds us that eating has long been imbued with religious and moral significance. One might think then that the naming of ON by Bratman in 1997 might have crystallized a form of suffering increasingly common in the health-obsessed West. After all, the imperative for the neoliberal subject to "eat right" predates Bratman's early publications (Crawford 1980; Biltekoff 2013; Cederström and Spicer 2015; Finn 2017). However, ON did not enter the realm of popular discourse following its formal introduction by Bratman, or even well after that. Only in the last decade has attention to it increased substantially. This assertion is supported by many blog posts, peer-reviewed literature, and over a hundred-thousand Instagram posts accompanied by the hashtag #orthorexia or similar terminology. Social media has provided a platform for lifestyle influencers, nutrition experts, and health gurus to share their wisdom with a larger audience than previously possible. Aside from simply informing consumers, this overabundance of information also reinforces the dominant perception that each individual is responsible for their own health, particularly in a context of neoliberal policies and ideologies. The sociologist Nicolosi (2016) has suggested that, in fact, we live in an orthorexic society, arguing that, "the individualisation of alimentary choices, reflecting a more general propensity toward autonomy from the social of the modern individual-consumer, opens up, alongside important degrees of freedom, distressing new forms of neophobia" (54).

What seems like "autonomy from the social" might instead manifest a new proliferation of sometimes contradictory alimentary discourses. Increasingly, many consumers resort to popularized diets or social media icons serve as a satisfactory proxy for inherited wisdom, fuelling in novel ways the adoption of what political economist Crawford (1980) called "healthism"—and subsequently, pushback against it, often in the form of other diet ideologies, such as the re-embrace of intuitive eating (Tribole and Resch 1995; Harrison 2019). The contestation amongst these sources and the increasing discourse on food in general foster an environment of intense examination of dietary regimes, as witnessed in the booming discussion of those that seem to go too far in an effort to achieve optimal health. While this self-centred embodiment of health values, the alienation between consumer and food, and exposure to often-contradictory nutritional information help explain why orthorexia nervosa is currently attracting so much attention, it is still uncertain at which point culturally valued behaviours become pathological.

13.4 Criteria in Context

Bratman himself and some of the other clinicians writing on the condition emphasize the importance of distress and impairment—either physiological, such as malnutrition, or psychosocial, such as fractured relationships or career challenges, resulting from orthorexia nervosa. The creators of ORT0-15 focused solely on the fixation and behaviours, according to Bratman and his co-author Dunn, potentially pathologising large numbers of "healthy" eaters (Dunn and Bratman 2016; Dunn et al. 2017). Given the contemporary imperative in the United States to "care for" one's body through food, which includes the employment of nutritional knowledge and often the identification of "good" and "bad" foods, some of the diagnostic criteria for ON reflect what are, to some degree, normalized, even valorised, approaches to food in American society (van Dyke 2018). For Bratman and Dunn, this requires clear criteria for delineating when health-promoting behaviours become pathological. Specifically, they stress the need to assess "clinically significant behavior, interpersonal distress, or medical problems concerning diet" (Dunn and Bratman 2016: 14). This call to focus on distress and impairment has been influential, but it is relatively recent; much of the published work on ON was conducted earlier, using what are now seen as deeply flawed modes of assessment, such as those mentioned earlier that find extremely high ON prevalence among some groups.

Yet even with the care of some researchers to not pathologise healthy eating, the line between lifestyle and disease is blurry. Dunn and Bratman (2016) include as diagnostic criterion: "Intrapersonal distress or impairment of social, academic, or vocational functioning secondary to beliefs or behaviors about healthy food" (16). Moroze et al. (2015) include a similar criterion. Such criteria hold the possibility of pitting what may actually be health-promoting behaviours, such as careful food preparation, against those that may be health-threatening, despite being deemed socially necessary, such as long work hours. Furthermore, to what degree does "social, academic, or vocational functioning" reflect particular social norms, to which some individuals may be held more than others?

Moroze et al. (2015) follow their list of criteria with what they call Criterion D: "The behavior is not better accounted for by the exclusive observation of organized orthodox religious food observance" (401, Table 2). It continues to exclude those pursuing special diets due to professionally diagnosed allergies or other conditions, but the religious food observance exclusion links group belief and practice to the normal, pathologising that which may *not* reflect socially recognized food choices. Such criterion has precedence; one of the DSM-5 criterion for ARFID is: "The disturbance is not better explained by lack of available food or *by an associated culturally sanctioned practice*" (APA 2013, emphasis added).

Related to this is the emphasis on time in some diagnostic assessments. For example, for Moroze et al. (2015) the proposed diagnosed criteria includes the following: "For individuals who are not food professionals, excessive amounts of time (e.g. 3 or more hours per day) spent reading about, acquiring, and preparing specific types of foods based on their perceived quality and composition" (401, Table 2). In research one of us has conducted over the last decade and a half in rural Mexico, three hours is a typical or even low amount of time spent on daily food labour. Much of this work is time consuming because of particular notions about what makes food good, both for taste and for health, and the two are not necessarily seen as mutually exclusive (Wynne 2020). To what degree is the time-based criterion for orthorexia based on normative notions of what is reasonable food labour in a social setting with easy access to convenience foods and long workdays?

Similarly, Moroze et al. (2015) propose as one criterion: "Spending excessive amounts of money relative to one's income on foods because of their perceived quality and composition" (401, Table 2). Again, this suggests that normativity is determined, in part, on the basis of one's status, in this case, financial status. How might such criterion pathologise poor or working-class people's efforts to eat healthfully, while the allocation of tremendous resources on the part of the wealthy, to, for example, hire personal chefs, may be deemed reasonable expenditures? While many of the psychologists examining ON have noted the sociocultural conditions necessary for its existence—stigmas attached to obesity; access to foods seen as healthy; and time and money for food labour, intellectual and manual—they tend to fall back on psychological traits to explain why some individuals may be more prone to the condition than others (e.g. Gleaves et al. 2013; Koven and Abry 2015; Turner and Lefevre 2017; McComb and Mills 2019). What the ON literature lacks is substantive and meaningful consideration of the ways in which the diagnostic criteria itself may pathologise some individuals' food activities and not others' depending on their social location.

Some of the diagnostic criteria and affiliated traits also reflect distinctly American notions of pleasure and health, most clearly in the assumption that the two are mutually exclusive. Central to Bratman's original book on ON, *Health FoodJunkies*, was the premise that orthorexics prioritized health over pleasure with regard to food (Bratman and Knight 2000). Although he still notes the importance of this, it is not part of his and Dunn's more recently proposed diagnostic criteria (Dunn and Bratman

2016). Challenging the dichotomization of pleasure and health is crucial. Scholarship on eating and pleasure in the last decade has made two powerful claims (simplified for our purposes here): one, that the pleasure of eating "unhealthy" foods may in fact promote a sort of short-term emotional health (e.g. Berlant 2010; Klein 2010); and two, that restrictive eating, whether or not it contributes to long-term health, may itself produce pleasure (e.g. O'Connor and Van Esterik 2008; Finn 2017). Throughout *Health FoodJunkies*, Bratman (2000) writes of the desirable feeling of lightness his restrictive diets gave him. Some diagnostic assessments note the "moral superiority" and "self-satisfaction" experienced by orthorexics (Varga et al. 2014; Koven and Abry 2015). Such assessments suggest that these pleasurable bodily sensations and affective states are themselves markers of pathology, but the agency and voices of orthorexics should not be discounted.

13.5 Recovering Orthorexics Speak

Where are the voices of orthorexics circulating accounts of lived experiences of the condition? ON figures prominently in social media and journalism, but extended works on it are limited. There are only four English-language print books on ON. The first published, *Health FoodJunkies*, has already been discussed throughout this paper, given the critical role of its author Stephen Bratman in the emergence of the concept of ON.[2] We focus our analysis here on two of these books, whose authors, like Bratman, represent the growing population who do identify with the ON label. It should be noted that while Bratman is formally educated in these matters, many of the authors that provide publicly accessible information on orthorexia, including the two whose texts we examine here, Jordan Younger and Edward Yuen, are not.

Out of the two, Jordan Younger undeniably has more influential power. Her 211,000 Instagram followers and thriving blog and podcast series attest to her social media prowess and outreach capabilities. This prominence was not achieved gradually over years, but occurred instantaneously after successfully selling over 40,000 copies of her cleansing program while briefly enrolled as a fiction-writing graduate student at the New School (Younger 2016). This success encouraged her to pursue her newfound passion as a social influencer full time, a pursuit which led to the overnight popularity of her best-selling book *Breaking Vegan.* This work reads as a narrative of Younger's life, from the perfectionist attitude she acquired in childhood to her decision to detach from strict vegan practices.

Edward Yuen, on the other hand, has not achieved such mainstream attention from his memoir *Beating Orthorexia*, which has only seven ratings on Amazon in comparison to Younger's 255. He boasts a meagre 144 followers on Twitter and out

[2] The fourth, *Orthorexia: When Healthy Eating Goes Bad*, was written by British dietician Renee McGregor (2017). Although McGregor supports many of Bratman's ideas and, like him, advocates for greater scientific precision in discussions of ON, she alone, of these four authors, does not identify as a recovering orthorexic in her text.

of his 18 blog posts only one has any reader comments. Like Younger, Yuen has no formal medical education, and writes from the perspective of a college student with the lone goal of helping any potential orthorexics who may need it. While his book is similar to Younger's in the auto-biographical sense, it is more heavily characterized by the cynical tone underlying his critique of fad diets and general nutritional knowledge. Although a less active voice in ON discourse, Yuen, who also goes by the social media handle "Health Freak Eddy", has positioned himself as a prominent lay expert on ON. Despite being self-published, his book *Beating Orthorexia* is significant for two reasons; first, given the lack of published books on ON, as of 2020, Yuen's text is near the top of the list in a Google or Amazon search, and second, it features the Bratman seal of approval, with a quote from the ON originator himself on the back cover. Yet, Yuen himself does not present a clear take on whether or not ON is a standalone disease.

For example, early on in his book, he writes of ON, "I don't think it's a disease, but if it's impacting your life negatively, that's enough reason to treat it seriously" (2017: 12–13). Later he argues, "Whether or not this affliction ever gets categorized as a mental 'disorder' doesn't matter. It is real, and it is holding people back" (Ibid.: 15). It is this impairment to which Yuen attaches his arguments most strongly, despite seeming ambivalence about the medicalisation of this impairment. Rooting his arguments in his own experience rather than in clinical encounters or formal research, Yuen challenges the assumption held by many professional researchers of ON that sufferers may have neurocognitive deficits. Rather he suggests that a common characteristic may be perfectionism, what he seems to see as a malleable personality trait that can be relaxed in the individual to reduce anxiety around food (Ibid.: 40). Surprisingly, his remedy is a series of eating guidelines, though he promises what he calls "The One Rule Diet"—Eat Real Food. Instead he advocates for everything from intermittent fasting to avoiding wheat bread.

In contrast, Jordan Younger, who is frequently mentioned as *the* influencer who brought ON to popular attention, clearly identifies herself as having an eating disorder throughout her book, published in 2016. In the forward to the book, titled *Breaking Vegan*, Stephen Bratman (2016) himself writes, "[T]the word *orthorexia* serves as a signifier. It is a kind of mental goalpost to indicate a limit, a boundary not to go beyond even in search of healthy diet. And, if you've already gone beyond, it can help you find your way back" (10). Mirroring this notion of a point at which pathological eating and thinking begins, Younger reflects on her own experience:

> Something I can also see with clarity now is that writing the cleanse and then running an email thread throughout the week for all the cleanse participants was one of the catalysts that tipped me over the edge from food obsession into full-blown orthorexia. (Ibid.: 67)

Younger describes herself as having an "addiction" to health eating, one that she connects at different points in the text to her "extreme personality", "crazy stomach problems", and her tendency to compare herself "to other people in the health community and the multitude of diets they follow and swear by" (Ibid.: 73). Yet social impairment appears in Younger's narrative far before she reaches what she calls "full-blown orthorexia". A recurring theme in the text is a lack of commensality and barriers to other forms of in-person social bonding—a dearth of romantic relationships, occasionally-strained friendships, and very few shared meals. Younger's obsession travels with her; she recounts packing a serrated knife in her carry-on before a flight so she could cut an avocado while on board. After the knife is detected by airport security, she begs them to at least leave her with the foodstuffs she has packed: almond milk and coconut yogurt, among them (Ibid.: 44–46).

While social impairment is evident through much of Younger's narrative, her experience poses challenges for applying some of the other proposed diagnostic criteria. As a food and lifestyle blogger, she might qualify as a food professional, in which case spending more than three hours a day focused on food or spending a good part of her income on food would not necessarily be seen as pathological. Her embeddedness in various, mostly virtual, food communities meant that many of her ideas and practices around food were normalized. It is when she abandons veganism and calls herself an orthorexic that she encounters intense criticism from many of her followers. While she details some of this in her book, its title also earned her vitriol; *Breaking Vegan*, according to some critics, equated veganism with orthorexia, perhaps even demonizing the former more than the latter. Yet the benefits of her new stance were clear: Younger became *the* spokesperson for ON in popular media, appearing on multiple major news programs and interviewed by dozens of publications.

In November 2018, Younger posted a new, and perhaps for her readers, shocking update on her blog. Explaining her recent decision to again embrace veganism in a blog post titled "From Breaking Vegan & Back Again: A Plant-Based Journey", she wrote that, more important than being what she described as "somewhat orthorexic":

> I was first and foremost legitimately sick — with extreme gastrointestinal issues, and what I now know were the deteriorating effects of mold poisoning and chronic Lyme disease. . . .So yes while I dealt with the tendencies of orthorexia back then as a means of sheer confusion about my health, I now know the root cause of my sickness was physical. I was sick, sick, sick with Lyme back then just as I am now. Orthorexia for me was a direct byproduct of desperately trying to heal myself in the only way I knew how. I don't resonate with much of what I believed about my health and orthorexic journey back then... because again, I was listening to people outside of myself (nutritionists, doctors, therapists) who also wanted to put me in another box: the eating disorder box. (2018)

ON's unofficial spokesperson is now distancing herself from the condition and again touting lifestyle recommendations on her social media accounts, including everything from coffee enemas to cupping to *ayahuasca*.

It would be easy to paint the authors we've examined as self-interested career healthists, more-or-less lay experts making a living or at least some income by touting their own carefully cultivated images and experiences. That may well be a fair criticism; at best, these authors simplify or ignore the most meaningful debates in the mental health literature on ON. Yet their public platforms, especially Younger's, have been influential in circulating the very notion of ON in popular discourse. In their books, the authors range from noncommittal to largely convinced as to whether ON is a disease, but their claiming of the condition to name their own experiences certainly plays a part in reifying it. In Yuen's questioning of ON as a disease and in

Younger's later questioning, in the 2018 blog post, whether she was orthorexic at all, these authors demonstrate that while they are not uncritical consumers of ON as a condition, for them, ON does in fact exist.

The DSM itself is cautious of the reification of its categories; as mentioned earlier, it encourages mental health professionals to think of it as a "*practical, functional, and flexible guide*" (APA 2013: Preface). While professional conferences and clinical settings are spaces in which that reification may unfold, regardless of the DSM's intent, we must also look at other discursive spaces in which mental disorders, including eating disorders like ON, become bounded entities. The tremendous proliferation of easily accessible diet-related information is often pointed to as a causal factor in the prevalence of ON, and yet those same informational structures now play a role in transforming ON into a culturally recognizable category.

In our critical analysis of the literature on ON, we do not aim to minimize the suffering of those with orthorexic symptoms. Rather we are attempting to complicate the sometimes overlapping, sometimes conflicting notions of health that emerge in debates about this condition. ON exposes and, in fact, embodies many of the contradictions of contemporary eating in the West. Perhaps the often incongruous and confused narratives of recovering orthorexics are themselves revealing of the challenge of distinguishing the virtuous from the diseased.

References

Alvarenga, Marie Dos Santos, Marcia Martins, Karen Sayuri Sato, Sílvia V.A. Vargas, Sonia Tucunduva Philippi, and Fernanda Baeza Scagliusi. 2012. Orthorexia nervosa behavior in a sample of Brazilian dietitians assessed by the Portuguese version of ORTO-15. *Eating and Weight Disorders—Studies on Anorexia, Bulimia and Obesity* 17(1): e29–e35. https://doi.org/10.1007/BF03325325.

American Psychiatric Association (APA). 2011. Frequently Asked Questions, Diagnostic and Statistical Manual of Mental Disorders. http://www.psych.org/MainMenu/Research/DSMIV/FAQs.aspx. Accessed 2 May 2020.

American Psychiatric Association (APA). 2013. *American Psychiatric Association: Diagnostic and Statistical Manual of Mental Disorders: Diagnostic and Statistical Manual of Mental Disorders*, 5th ed. Arlington, VA: American Psychiatric Association.

Berlant, Lauren. 2010. Risky Bigness: On Obesity, Eating, and the Ambiguity of "Health." In *Against Health: How Health Became the New Morality*, ed. Jonathan M. Metzl and Anna Kirkland, 26–39. New York: New York University Press.

Biltekoff, Charlotte. 2013. *Eating Right in America: The Cultural Politics of Food & Health*. Durham and London: Duke University Press.

Bratman, Steven. 1997. Original essay on orthorexia. http://www.orthorexia.com/original-orthorexia-essay/. Accessed 30 Sept 2019.

Bratman, Steven, and David Knight. 2000. *Health food junkies. Orthorexia nervosa: Overcoming the obsession with healthful eating*. New York: Random House.

Cederström, Carl, and André Spicer. 2015. *The Wellness Syndrome*. Cambridge: Polity Press.

Cena, Hellas, Friederike Barthels, Massimo Cuzzolaro, Steven Bratman, Anna Brytek-Matera, Thomas Dunn, Marta Varga, Benjamin Missbach, and Lorenzo M. Donini. 2018. Definition and diagnostic criteria for orthorexia nervosa: A narrative review of the literature. *Eating and*

Weight Disorders—Studies on Anorexia, Bulimia and Obesity. https://doi.org/10.1007/s40519-018-0606-y.

Cosgrove, Lisa, and Sheldon Krimsky. 2012. A Comparison of *DSM*-IV and *DSM*-5 Panel Members' Financial Associations with Industry: A Pernicious Problem Persists. *PLoS Medicine* 9 (3): e1001190. https://doi.org/10.1371/journal.pmed.1001190.

Crawford, Robert. 1980. Healthism and the medicalization of everyday life. *International Journal of Health Services* 10 (3): 365–388.

Donini, Lorenzo M., Daniela Marsili, Maria Pia Graziani, Michele Imbriale, and Carlo Cannella. 2004. Orthorexia nervosa: A preliminary study with a proposal for diagnosis and an attempt to measure the dimension of the phenomenon. *Eating and Weight Disorders—Studies on Anorexia, Bulimia and Obesity* 9: 151–157. https://doi.org/10.1007/BF03325060.

Donini, Lorenzo M., Daniela Marsili, Maria Pia Graziani, Michele Imbriale, and Carlo Cannella. 2005. Orthorexia nervosa: Validation of a diagnosis questionnaire. *Eating and Weight Disorders—Studies on Anorexia, Bulimia and Obesity* 10(2): e28–e32. https://doi.org/10.1007/bf03327537.

Dunn, Thomas M., and Steven Bratman. 2016. On orthorexia nervosa: A review of the literature and proposed diagnostic criteria. *Eating Behaviors* 21: 11–17. https://doi.org/10.1016/j.eatbeh.2015.12.006.

Dunn, Thomas M., Josh Gibbs, Noelle Whitney, and Amy J. Starosta. 2017. Prevalence of orthorexia nervosa is less than 1%: Data from a US sample . *Eating and Weight Disorders—Studies on Anorexia, Bulimia and Obesity* 22: 185–192. https://doi.org/10.1007/s40519-016-0258-8.

Finn, S. Margot. 2017. *Discriminating Taste: How Class Anxiety Created the American Food Revolution*. New Brunswick, Camden, and Newark, New Jersey, and London: Rutgers University Press.

Friedman, Richard A. 2013. The Book Stops Here. *The New York Times*, May 20. https://www.nytimes.com/2013/05/21/health/the-dsm-5-as-a-guide-not-a-bible.html.

Gleaves, David H., Erin C. Graham, and Suman Ambwani. 2013. Measuring "orthorexia": Development of the Eating Habits Questionnaire. *The International Journal of Educational and Psychological Assessment* 12 (2): 1–18.

Greenberg, Gary. 2013. *The Book of Woe*. New York, NY: Blue Rider Press.

Håman, Linn, Natalie Barker-Ruchti, Göran Patriksson, and Eva-Carin Lindgren. 2015. Orthorexia nervosa: an integrative literature review of a lifestyle syndrome. *International Journal of Qualitative Studies of Health and Well-being* 10. https://doi.org/10.3402/qhw.v10.26799

Harrison, Christy. 2019. *Anti-Diet: Reclaim Your Time, Money, Well-Being, and Happiness Through Intuitive Eating*. New York, Boston, London: Little, Brown Spark.

Kawa, Shadia, and James Giordano. 2012. A brief historicity of the Diagnostic and Statistical Manual of Mental Disorders: issues and implications for the future of psychiatric canon and practice. *Philosophy, Ethics, and Humanities in Medicine*. https://doi.org/10.1186/1747-5341-7-2.

Klein, Richard. 2010. What is Health and How Do You Get It? In *Against Health: How Health Became the New Morality*, ed. Jonathan M. Metzl and Anna Kirkland, 15–25. New York: New York University Press.

Koven, Nancy S., and Alexandra W. Abry. 2015. The clinical basis of orthorexia nervosa: Emerging perspectives. *Neuropsychiatric Disease and Treatment* 11: 385–394. https://doi.org/10.2147/NDT.S61665.

Lindeman, Maarjana, Pertti Keskivaara, and Maarit Roschier. 2000. Assessment of magical beliefs about food and health. *Journal of Health Psychology* 5 (2): 195–209. https://doi.org/10.1177/135910530000500210.

McComb, Sarah E., and Jennifer S. Mills. 2019. Orthorexia nervosa: A review of psychosocial risk factors. *Appetite* 140: 50–75. https://doi.org/10.1016/j.appet.2019.05.005.

McGregor, Renee. 2017. *Orthorexia: When Healthy Eating Goes Bad*. London: Watkins.

Missbach, Benjamin, Barbara Hinterbuchinger, Verena Dreiseitl, Silvia Zellhofer, Carina Kurz, and Jürgen König. 2015. When Eating Right, Is Measured Wrong! A Validation and Critical Examination of the ORTO-15 Questionnaire in German. *PloS one* 10(8). https://doi.org/10.1371/journal.pone.0135772.

Moroze, Ryan M., Thomas M. Dunn, J. Craig Holland, Joel Yager, and Philippe Weintraub. 2015. Microthinking about micronutrients: A case of transition from obsessions about healthy eating to near-fatal "orthorexia nervosa" and proposed diagnostic criteria. *Psychosomatics* 56 (4): 397–403. https://doi.org/10.1016/j.psym.2014.03.003.

Nicolosi, Guido. 2006. Biotechnologies, alimentary fears and the orthorexic society. *Tailoring Biotechnologies* 2 (3): 37–56.

O'Connor, Richard, and Penny Van Esterik. 2008. De-medicalizing anorexia: A new cultural brokering. *Anthropology Today* 24: 6–9.

Ramacciotti, Carla E., Patrícia Perrone, Elisabetta Coli, Annalisa Burgalassi, Ciro Conversano, Gabriele Massimetti, and Liliana Dell'Osso. 2011. Orthorexia nervosa in the general population: A preliminary screening using a self-administered questionnaire (ORTO-15). *Eating and Weight Disorders—Studies on Anorexia, Bulimia and Obesity* 16(2): e127–e130. https://doi.org/10.1007/bf03325318.

Tribole, Evelyn, and Elyse Resch. 1995. *Intuitive Eating: A Recovery Book for the Chronic Dieter: Rediscover the Pleasures of Eating and Rebuild Your Body Image*. London: St. Martin's Press.

Turner, Pixie G., and Carmen E. Lefevre. 2017. Instagram use is linked to increased symptoms of orthorexia nervosa . *Eating and Weight Disorders—Studies on Anorexia, Bulimia and Obesity* 22: 277–284. https://doi.org/10.1007/s40519-017-0364-2.

Van Dyke, Christina. 2018. Eat y'self Fitter: Orthorexia, Health, and Gender. In *The Oxford Handbook of Food Ethics*, ed. Anne Barnhill, Mark Budolfson, and Tyler Doggett, 553–571. New York: Oxford University Press.

Varga, Márta., Barna Konkolÿ Thege, Szilvia Dukay-Szabó, F. Ferenc Túry, and Eric F. van Furth. 2014. When eating healthy is not healthy: Orthorexia nervosa and its measurement with the ORTO-15 in Hungary. *BMC Psychiatry* 14 (1): 59–70. https://doi.org/10.1186/1471-244X-14-59.

Wynne, Lauren A. 2020. *Predictable Pleasures: Food and the Pursuit of Balance in Rural Yucatán*. Lincoln, NE: University of Nebraska Press.

Younger, Jordan. 2016. *Breaking Vegan: One Woman's Journey from Veganism, Extreme Dieting, and Orthorexia to a More Balanced Life*. Beverly, MA: Fair Winds Press.

Younger, Jordan. 2018. From Breaking Vegan & Back Again: A Plant-Based Journey. *The Balanced Blonde*, https://www.thebalancedblonde.com/2018/11/26/from-breaking-vegan-back-again-a-plant-based-journey/. Accessed 1 October 2019.

Yuen, Edward. 2015. *Beating Orthorexia and the Memoirs of a Health Freak*. Self-published.

Lauren A. Wynne is an Assistant Professor of Anthropology in the Department of Anthropology and Sociology of Ursinus College in Pennsylvania. She holds a Ph.D. in Sociocultural Anthropology from The University of Chicago. She is a cultural and medical anthropologist who has conducted work on food practices in rural Yucatán, Mexico, and on pregnancy and childbirth experiences in New York. She is the author of *Predictable Pleasures*: *Foodand the Pursuit of Balance in Rural Yucatán*, published in 2020.

Gareen Hamalian is a clinical assistant professor in the Department of Psychiatry at New York University School of Medicine. After training in general psychiatry and completing a fellowship in forensic psychiatry at New York University School of Medicine, she has worked in a variety of settings including inpatient, outpatient, integrative care and as a consultant and forensic evaluator. Her research and teaching interests include nutritional psychiatry, telemedicine, reproductive psychiatry and the impact of dermatological illness on mental health.

Neve Durrwachter is a graduate student in public health at Johns Hopkins University. Her research interests include historical conceptions of health and diet as well as the burgeoning discourse on orthorexia nervosa.

Index

S. Stano and A. Bentley (eds.), *Food for Thought*, Numanities - Arts and Humanities in Progress 19, https://doi.org/10.1007/978-3-030-81115-0

GPSR Compliance

The European Union's (EU) General Product Safety Regulation (GPSR) is a set of rules that requires consumer products to be safe and our obligations to ensure this.

If you have any concerns about our products, you can contact us on ProductSafety@springernature.com

In case Publisher is established outside the EU, the EU authorized representative is:

Springer Nature Customer Service Center GmbH
Europaplatz 3
69115 Heidelberg, Germany

Batch number: 10756376

Printed by Printforce, the Netherlands